LOW-VOLTAGE CMOS LOG COMPANDING ANALOG DESIGN

THE KLUWER INTERNATIONAL SERIES IN ENGINEERING AND COMPUTER SCIENCE

ANALOG CIRCUITS AND SIGNAL PROCESSING

Consulting Editor*: Mohammed Ismail. *Ohio State University

Related Titles:

SYSTEMATIC DESIGN OF ANALOG IP BLOCKS
Vandenbussche and Gielen
ISBN: 1-4020-7471-9

SYSTEMATIC DESIGN OF ANALOG IP BLOCKS
Cheung & Luong
ISBN: 1-4020-7466-2

LOW-VOLTAGE CMOS LOG COMPANDING ANALOG DESIGN
Serra-Graells, Rueda & Huertas
ISBN: 1-4020-7445-X

CIRCUIT DESIGN FOR WIRELESS COMMUNICATIONS
Pun, Franca & Leme
ISBN: 1-4020-7415-8

DESIGN OF LOW-PHASE CMOS FRACTIONAL-N SYNTHESIZERS
DeMuer & Steyaert
ISBN: 1-4020-7387-9

MODULAR LOW-POWER, HIGH SPEED CMOS ANALOG-TO-DIGITAL CONVERTER FOR EMBEDDED SYSTEMS
Lin, Kemna & Hosticka
ISBN: 1-4020-7380-1

DESIGN CRITERIA FOR LOW DISTORTION IN FEEDBACK OPAMP CIRCUITE
Hernes & Saether
ISBN: 1-4020-7356-9

CIRCUIT TECHNIQUES FOR LOW-VOLTAGE AND HIGH-SPEED A/D CONVERTERS
Walteri
ISBN: 1-4020-7244-9

DESIGN OF HIGH-PERFORMANCE CMOS VOLTAGE CONTROLLED OSCILLATORS
Dai and Harjani
ISBN: 1-4020-7238-4

CMOS CIRCUIT DESIGN FOR RF SENSORS
Gudnason and Bruun
ISBN: 1-4020-7127-2

ARCHITECTURES FOR RF FREQUENCY SYNTHESIZERS
Vaucher
ISBN: 1-4020-7120-5

THE PIEZOJUNCTION EFFECT IN SILICON INTEGRATED CIRCUITS AND SENSORS
Fruett and Meijer
ISBN: 1-4020-7053-5

CMOS CURRENT AMPLIFIERS; SPEED VERSUS NONLINEARITY
Koli and Halonen
ISBN: 1-4020-7045-4

MULTI-STANDARD CMOS WIRELESS RECEIVERS
Li and Ismail
ISBN: 1-4020-7032-2

A DESIGN AND SYNTHESIS ENVIRONMENT FOR ANALOG INTEGRATED CIRCUITS
Van der Plas, Gielen and Sansen
ISBN: 0-7923-7697-8

RF CMOS POWER AMPLIFIERS: THEORY, DESIGN AND IMPLEMENTATION
Hella and Ismail
ISBN: 0-7923-7628-5

DATA CONVERTERS FOR WIRELESS STANDARDS
C. Shi and M. Ismail
ISBN: 0-7923-7623-4

DIRECT CONVERSION RECEIVERS IN WIDE-BAND SYSTEMS
A. Parssinen
ISBN: 0-7923-7607-2

AUTOMATIC CALIBRATION OF MODULATED FREQUENCY SYNTHESIZERS
D. McMahill
ISBN: 0-7923-7589-0

MODEL ENGINEERING IN MIXED-SIGNAL CIRCUIT DESIGN
S. Huss
ISBN: 0-7923-7598-X

ANALOG DESIGN FOR CMOS VLSI SYSTEMS
F. Maloberti
ISBN: 0-7923-7550-5

LOW-VOLTAGE CMOS LOG COMPANDING ANALOG DESIGN

by

Francisco Serra-Graells

Instituto de Microelectrónica de Barcelona, IMB-CNM

Adoración Rueda

Instituto de Microelectrónica de Sevilla-CNM

and

José L. Huertas

Instituto de Microelectrónica de Sevilla-CNM

KLUWER ACADEMIC PUBLISHERS

BOSTON / DORDRECHT / LONDON

A C.I.P. Catalogue record for this book is available from the Library of Congress.

ISBN 978-1-4419-5353-7 e-ISBN 978-0-306-48721-7

Published by Kluwer Academic Publishers,
P.O. Box 17, 3300 AA Dordrecht, The Netherlands.

Sold and distributed in North, Central and South America
by Kluwer Academic Publishers,
101 Philip Drive, Norwell, MA 02061, U.S.A.

In all other countries, sold and distributed
by Kluwer Academic Publishers,
P.O. Box 322, 3300 AH Dordrecht, The Netherlands.

Printed on acid-free paper

All Rights Reserved
© 2003 Kluwer Academic Publishers, Boston
Softcover reprint of the hardcover 1st edition 2003
No part of this work may be reproduced, stored in a retrieval system, or transmitted
in any form or by any means, electronic, mechanical, photocopying, microfilming, recording
or otherwise, without written permission from the Publisher, with the exception
of any material supplied specifically for the purpose of being entered
and executed on a computer system, for exclusive use by the purchaser of the work.

Contents

List of Figures

List of Tables

Acknowledgments

The authors wish to make public acknowledgment of the help from both the Centro Nacional de Microelectrónica and Microson S.A. staffs. Special thanks for the NEXO©-team: Lluís Gómez, Òscar Farrés, Ferran Casas and Xavier Aresté.

This work has been partially funded by European Community ESPRIT FUSE 23068 (Microelectronic Device for Hearing Aid Application) and Spanish CICYT TIC97-1159 (*Microelectrónica de Bajo Consumo y Baja Alimentación para Audífonos*) and CICYT TIC99-1084 (*Técnicas de Circuitos CMOS VLSI para Subsistemas Analógicos en Audífonos Digitales*) projects.

Acknowledgments

To my wife

Lurdes,

and to my sons

Marc and Anna.

F. Serra-Graells

Abstract This book presents in detail state-of-the-art analog circuit techniques for the very low-voltage and low-power design of systems-on-chip in CMOS technologies. The proposed strategy is mainly based on two basis: the Instantaneous Log Companding Theory, and the MOSFET operating in the subthreshold region. The former allows inner compression of the voltage dynamic-range for very low-voltage operation, while the latter is compatible with CMOS technologies and suitable for low-power circuits. In this sense, the required background on the specific modeling of the MOS transistor for Companding is supplied at the beginning.

Following this general approach, a complete set of CMOS basic building blocks are proposed and analyzed for a wide variety of analog signal processing. In particular, the covered areas include: amplification and AGC, arbitrary filtering, PTAT generation, and pulse duration modulation (PDM). For each topic, several case studies are considered to illustrate the design methodology. Also, integrated examples in 1.2μm and 0.35μm CMOS technologies are reported to verify the good agreement between design equations and experimental data. The resulting analog circuit topologies exhibit very low-voltage (i.e. 1V) and low-power (few tenths of μA) capabilities.

Apart from these specific design examples, a real industrial application in the field of hearing aids is also presented as the main demonstrator of all the proposed basic building blocks. This system-on-chip exhibits true 1V operation, high flexibility through digital programmability and very low-power consumption (about 300μA including the Class-D amplifier). As a result, the reported ASIC can meet the specifications of a complete family of common hearing aid models.

In conclusion, this book is addressed to both, industry ASIC designers who can apply its contents to the synthesis of very low-power systems-on-chip in standard CMOS technologies, as well as to the teachers of modern circuit design in electronic engineering.

Chapter 1

INTRODUCTION

Abstract This chapter is devoted to introducing the context of the presented work. Motivations for CMOS implementations of low-power system-on-chip applications are explained first. Then, limitations of the state-of-the-art CMOS analog techniques are studied. As a result, the Instantaneous Companding Theory is chosen here as the most efficient and complete signal processing alternative for very low-voltage operation. A short overview of this theory is presented through a generalized device-independent nomenclature. Problems on migration from previous bipolar implementations of this theory to modern CMOS processes are argued. Hence, the final goal of this work is defined as the research on novel CMOS analog circuit techniques compatible with CMOS technologies, which should exploit the low-voltage capabilities of Log Companding processing.

1. Low-Power Applications and CMOS Technologies

Portable and miniaturized system-on-chip applications have always exhibit an increasing demand in the microelectronics market and, particularly, in the biomedical field with products such as hearing aids, pacemakers or implantable sensors.

System portability usually requires battery supply, except in some special cases such as RF-powered telemetry systems. Unfortunately, battery technologies do not evolve as fast as applications demand, so the combination of battery supply and miniaturization often turns into a low-voltage and/or low-current circuit design problem. In particular, these restrictions affect more drastically the analog part of the whole mixed system-on-chip. As a result, specific analog circuit techniques are needed to cope with such power supply limitations.

Since analog design techniques are usually device-dependent, the choice of the most suitable technology is of basic importance. In this sense, fully integrated CMOS implementations are preferred to bipolar and BiCMOS approaches. Motivations for such a choice are mainly based on the expected evolution of the semiconductor technology. Predictions extracted from [1] argue that designs, thus, circuit techniques based on CMOS technologies will feature the following advantages:

Low Costs. In case of not requiring the last generation of sub-micron CMOS processes, like the CMOS analog circuit techniques proposed in this work, silicon area costs have become affordable even for small series of full-custom ASIC designs.

Mixed A/D. Today's semiconductor industry is mainly pushed by CMOS digital designs which actually set the specifications for the next generations of ULSI technologies. Any analog circuit technique capable of being compatible with future digital standards concerning supply voltage, as shown in Figure 1.1, and current consumption, will have greater chances of success in ever more digital environment.

Design Portability. It seems clear that analog designs using just the small set of standard CMOS devices (i.e. complementary MOSFETs and eventually poly-Si capacitors) must be easier to translate to other similar technologies than circuits requiring many different types of basic elements (e.g. complementary BJTs, rectifying diodes, JFETs, high-value passive resistors and capacitors or zener diodes), which must satisfy many electrical specifications.

μSystems. Current progress in N/MEMS [2] can make monolithic systems with smart sensors possible in the short term. The integration of both transducers and circuitry in the same silicon bulk could increase system performances and reduce the overall package size.

Large Scaling. Although every technological step forward has been preceded by a prediction of insoluble physical barriers, CMOS technologies always break their own scaling limitations and stand as the leader in the microelectronics semiconductor industry. This tendency is not expected to change in the next future.

In conclusion, research on low-power CMOS analog circuit techniques seems to be of particular interest for the current market demands as they combine both a standard technology and an increasing range of application products.

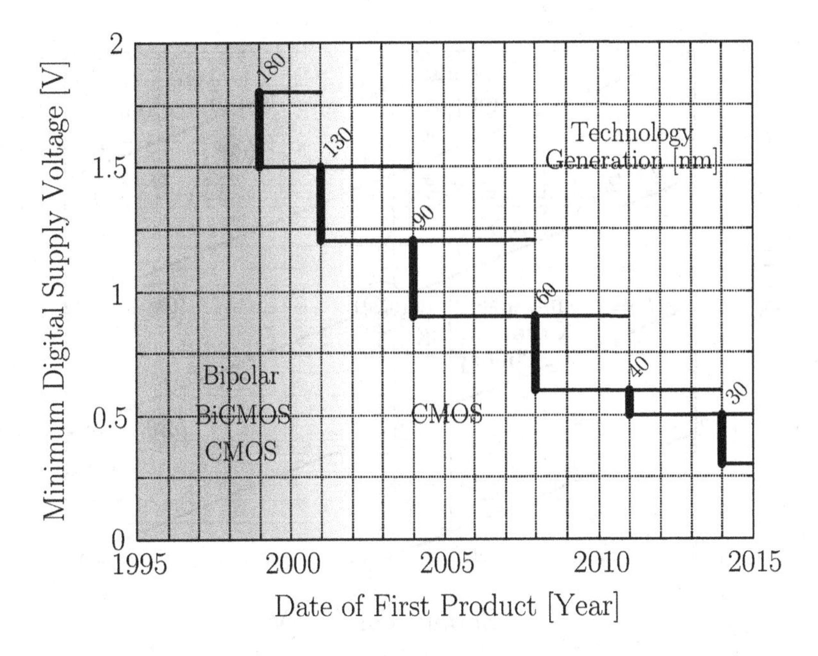

Figure 1.1. Prediction of Digital supply voltage scaling for ULSI technologies.

2. State-of-the-Art Low-Power Analog Design

The low-power term is often applied to both low-voltage and low-current circuit specifications without much discrimination. In fact, the circuit strategies adopted to overcome such design problems can even be in opposition. For example, specific biasing control techniques to deal with the scaling down of supply voltage may cause an increase in the overall current consumption due to the extra auxiliary circuitry added. Hence, it is important at this point to separate these two different design constraints in order to compare correctly the solutions existent in literature. The map of Figure 1.2 symbolizes this independence between low-voltage and low-current optimization strategies. The graphical comparison of the state-of-the-art analog circuit techniques refers to an arbitrary circuit design of $+5V/100\mu A$. A short description of the specific circuit approaches for low-voltage operation is listed below:

Rail-to-Rail includes all strategies oriented to extending the signal voltage range up to the available room between supply rails. Most of them are mainly based on the redesign of the input and output stages in order to increase their linear range [3, 4, 5, 6, 7].

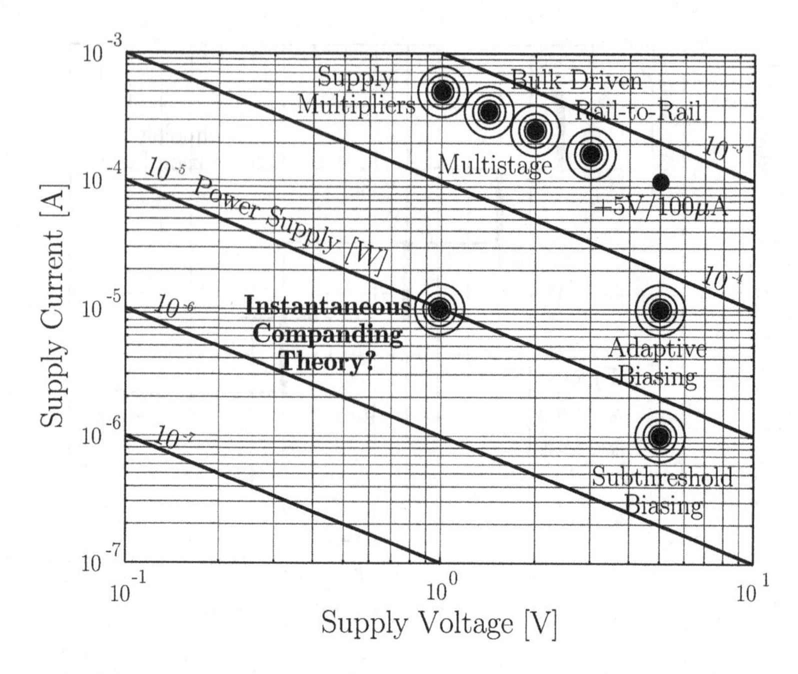

Figure 1.2. Map of state-of-the-art CMOS low-power analog continuous-time circuit techniques.

Multistage stands for multiple but simple cascaded stages instead of single cascoded structures. Efforts are then focused on their frequency stabilization with nested compensating loops [8, 9].

Bulk-Driven strategies make use of the MOSFET local substrate as an active signal terminal to obtain lower equivalent threshold voltages [10, 11].

Supply Multipliers bypass the low-voltage restriction by performing an step-up conversion of supply voltage through charge pumps [12, 13, 14, 15, 16, 17, 18, 19], typically from 1.5V to 3V.

In a similar way, the main circuit techniques for low-current consumption applications are enumerated as follows:

Adaptive Biasing is based on non-static current bias to optimize consumption according to signal demands. Bias dynamics are defined either by local positive feedback [20, 21] or by feedforward [22, 23] controls.

Subthreshold Biasing of classic topologies by operating their MOS transistors in the weak inversion region at very low-current levels [24].

The present work is focused on the low-voltage environment. In this case, a more detailed revision of the first circuit techniques reports the following drawbacks:

- All the low-voltage strategies except those using supply multipliers are actually partial solutions since they are addressed mainly to the design of operational amplifiers only.
- The bulk-driven option is also in opposition to general anti-latch-up rules of any standard CMOS process.
- Although supply multipliers are the only global and perhaps the most used solution for very low-voltage operation, they need large capacitive components, take an important Si area overhead and exhibit high extra current consumption, which make them not suitable for small package and low-current applications.

As a result, the next section introduces an alternative signal processing approach more suitable for low-voltage environments.

3. Instantaneous Companding Theory

Classic continuous-time analog signal processing [4] makes use of current (I) and voltage (V) as linear representations of internal signals. Practical integrated circuits implementing such processing usually appeal to passive components for the linear I/V behavior, while semiconductor devices are only devoted to auxiliary control, like operational amplifiers in active-RC techniques [26]. Even in fully integrated solutions, such as MOSFET-C and G_m-C [27], the active components are linearized to emulate this ideal linear law between I and V.

On the other hand, the aim of the companding signal processing [28, 29, 30, 31, 32] reviewed here is to exploit the intrinsic non-linear I/V characteristics of semiconductor devices to process signals linearly more efficiently. The basic idea of companding theories is to choose the I-domain for the input and output linear signals, with a given dynamic range (DR_I) defined according to:

$$DR[\text{dB}] \doteq 10 \log \left(\frac{\text{Maximum Input signal Power}}{\text{Minimum Input signal Power}} \right) \tag{1.1}$$

but yo process them internally using an equivalent V-domain with compressed dynamic range ($DR_V < DR_I$). The general scenario is depicted in Figure 1.3.

The first step in the signal chain includes an extra $I \rightarrow V$ compression and is ended by a $V \rightarrow I$ expansion of the original dynamic range, which

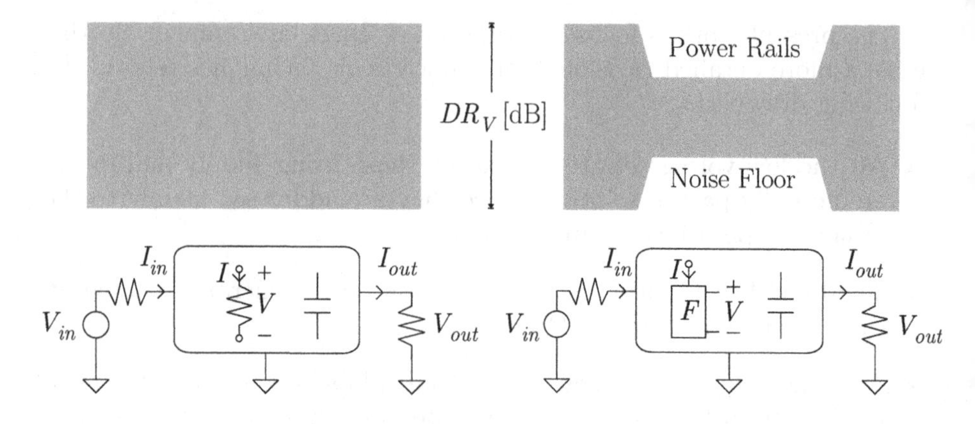

Figure 1.3. DR comparison between classic (left) and companding (right) processing schemes.

give the general name to the theory: companding=compressing+expanding. Since DR_I is kept constant through all these stages, voltage companding is also called the current-mode approach. However, the current-mode domain term [33] also includes other linear processing techniques like current conveyors-RC [34], so the first terminology is preferred here. The direct benefits of such internal non-linear processing are more tangible in the following environments:

High-Frequency. Since electric parasitic elements in any semiconductor planar technology are mainly capacitive, a reduction in the internal V dynamic range also decreases the portion of power wasted to charge and discharge the parasitic elements of the integrated circuit. Hence, larger bandwidth-to-power ratios can be achieved with the same devices. This was the original motivation of first companding signal processing theories [3].

Low-Voltage. Thanks again to the compression of the internal DR_V, large voltage swings around the bias point are reduced, so the necessary room between power rails can be scaled down without extra circuitry.

In order to obtain the desired internal DR_V, the companding processing is performed by basic building blocks with I/V characteristics similar to the general function depicted in Figure 1.4. Due to the curved shape, around any arbitrary amplitude reference (I_o), larger I signals will be always attenuated in the V-domain with respect to the equivalent reference (V_o), while lower I amplitudes will be amplified, all in a

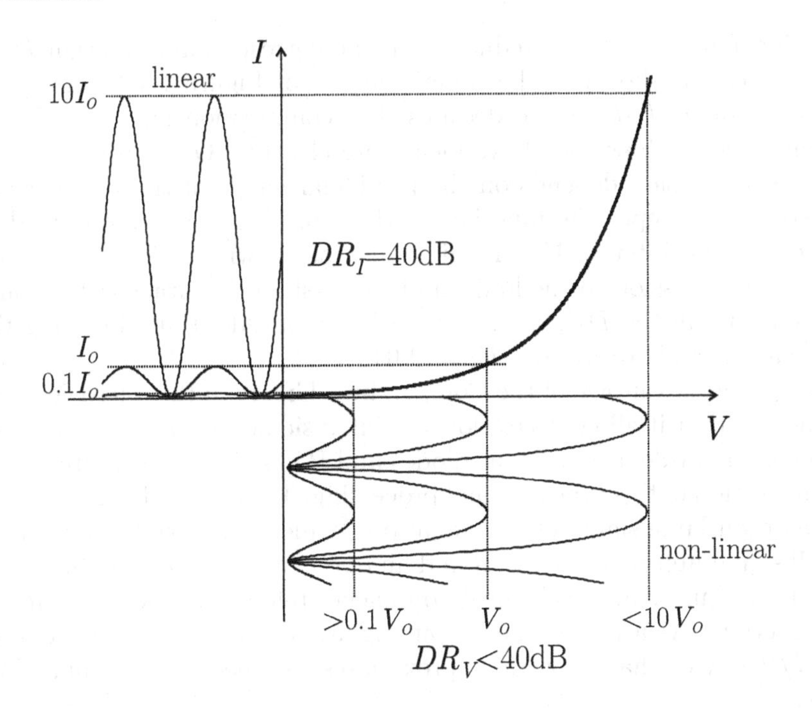

Figure 1.4. Example of DR_V compression for a $DR_I = 40$dB.

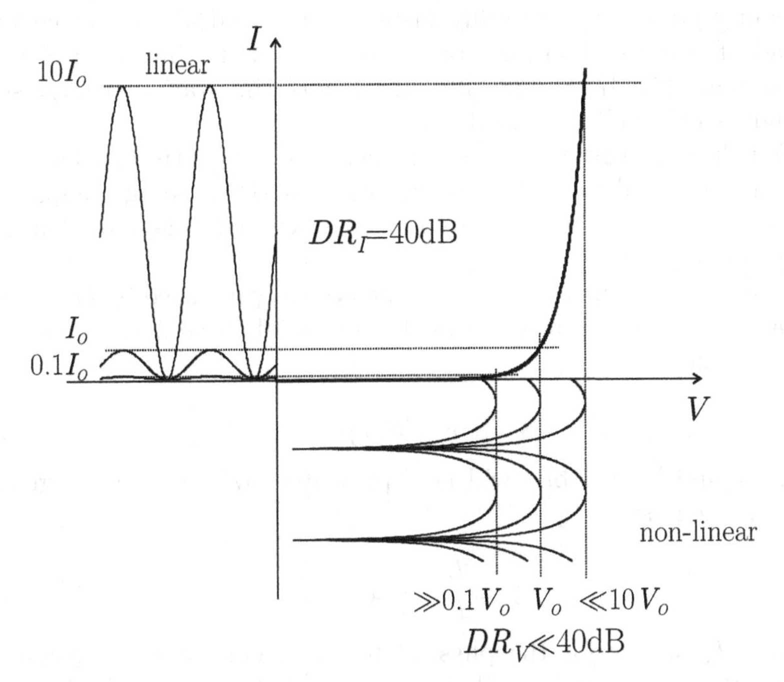

Figure 1.5. Example of a higher DR_V compression than that in Figure 1.4

non-linear way. Hence, a reduction in the corresponding internal DR_V is achieved, as shown in the same example of Figure 1.4 for an input dynamic range DR_I of two decades (low compression has been chosen in this case to show the above idea more clearly). Obviously, the final DR_V compression depends on the particular shape of the I/V characteristic: the steeper the function is, the more reduction is obtained in the equivalent internal V dynamic range, as shown in Figure 1.5. Taking this progression to the limit, an ideal ⅃-shaped characteristic would concentrate all the DR_I in the single V_o amplitude, thus shrinking the internal dynamic range to $DR_V = 0$dB.

The great advantage of the Companding Theory in the microelectronics field is that it allows to exploit the intrinsic non-linear I/V curves of semiconductor devices, such as diodes and BJTs, for a direct circuit implementation of this type of signal processing. Hence, the design strategy is based on large signal device equations, which has nothing to do with small signal approximations around an operating point or any linearization technique as in the classical approaches referred to the beginning of this section. As a result, the output signal is distortion-free as long as the I/V device characteristic approximates the theoretical companding curve.

It is important at this point to note that both the compressing and expanding processes are ideally instantaneous and at the device-level, instead of other system-level processing techniques including Syllabic companding [20, 37, 38, 39, 40, 21, 22] and time-variant compressing techniques for AGC such as [3, 4].

After the above overview, the nomenclature to be introduced next will allow a device-independent generalization of the existing companding theories, a necessary previous step for any new technology implementation proposal.

Actual semiconductor devices can be typically modeled by I/V curves similar to those in Figures 1.4 and 1.5 and symbolized here by the companding function F:

$$y = F(x) \tag{1.2}$$

Also I and V variables will be represented by the general signals y and x defined as:

$$y \propto \frac{I_k}{I_S} \qquad x \propto \frac{V_{ij}}{U_t} \tag{1.3}$$

where I_k stands for the physical terminal current (e.g. Anode or Collector), and I_S corresponds to some specific current including technological, geometrical and thermal parameters. In an equivalent way, V_{ij}

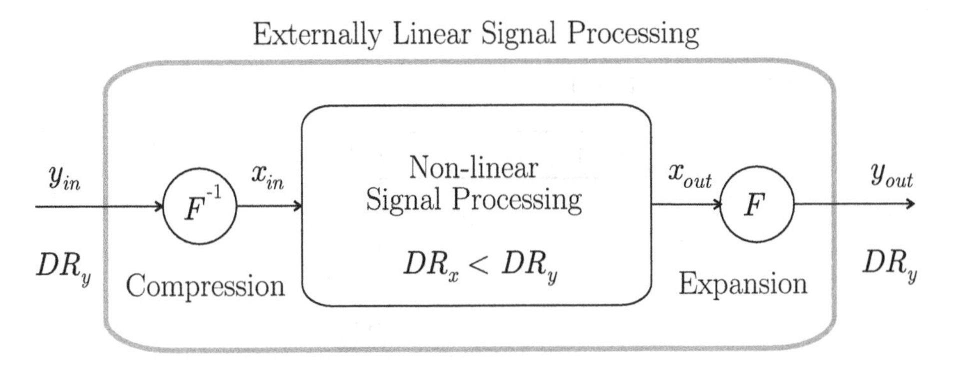

Figure 1.6. Signal domains for the generalized companding processing.

symbolizes the differential voltage between device terminals (e.g. Anode-Cathode or Base-Emitter) normalized to the thermal potential (U_t). In fact, the specific constants of proportionality in (1.3) depend on each particular semiconductor device.

From a processing point of view, the previous $y \to x$ compression stage can be now formulated as $x = F^{-1}(y)$ and the posterior $x \to y$ expansion as $y = F(x)$, both depicted in Figure 1.6. Although these pre and post stages are clearly non-linear, the key to any Companding Theory is to find the suitable internal x-domain processing for each F function, allowing a final cancellation of such non-linearities which may cause the system to be externally seen as linear.

Examples of internal DR_x compression can be computed for realistic F functions such as quadratic and exponential laws which give the name to Square-root and Log voltage companding, respectively. Results are shown analytically in Table 1.1 and graphically in Figure 1.7. In both cases the x-domain exhibits an important dynamic range reduction compared to the classical linear approach with ($DR_x \equiv DR_y$). Here it can be seen again that the total amount of reduction in DR_x depends on the specific shape of F. In the case of an exponential law, compression also changes along the region of operation defined by the bias point (x_o). Typical values of $V_{ij} > 0.4\text{V}$ at room temperature in most semiconductor devices make it necessary to locate $x_o > 16$. In these cases, DR_x compression tends to increase proportionally to this bias point according to:

$$DR_x = 20 \log \left(\frac{DR_y}{20 x_o \log(e)} + 1 \right) \simeq \frac{DR_y}{x_o} \qquad DR_y \ll 120\text{dB} \qquad (1.4)$$

Table 1.1. DR_x versus DR_y and F.

$F(x)$	DR_x
x	DR_y
x^2	$\frac{DR_y}{2}$
e^x	$\frac{DR_y}{x_o}$

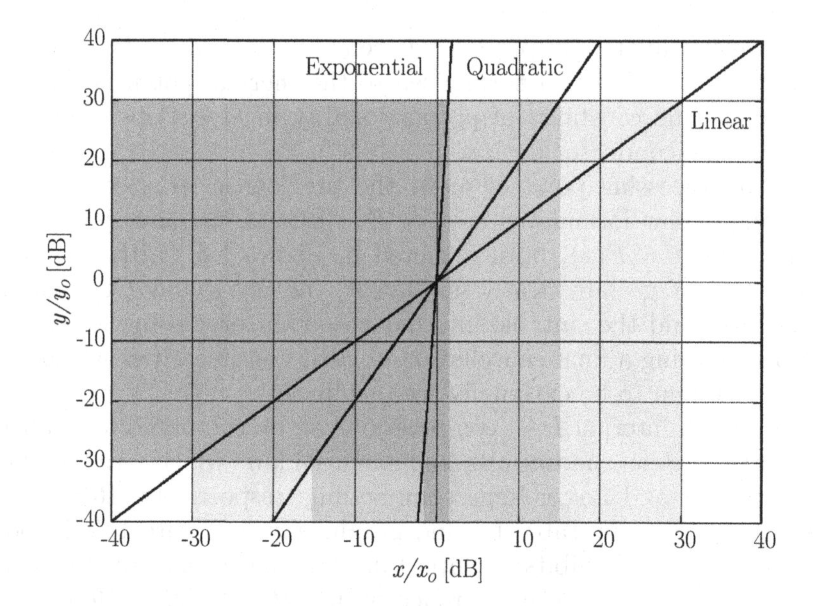

Figure 1.7. DR_x comparison between linear, quadratic and exponential laws (x_o = 20) companding functions F for DR_y = 60dB.

where both DR_x and DR_y are expressed in dB. As an example, the significant compression observed in Figure 1.7 for the Log case (less than 1 octave of DR_x per 3 decades of DR_y) agrees with the general idea that a forward-biased diode exhibits an almost constant differential voltage $V_{ij} \doteq V_{\text{Anode-Cathode}}$, never doubled or halved even for a large dynamic range of current $I_k \doteq I_{\text{Anode}}$.

4. CMOS Subthreshold Companding Proposal

From the Translinear Principle [4] to Log filters [1], most of the companding processing theories have chosen Log compression through exponential law F functions:

$$y = F(x) = e^x > 0 \tag{1.5}$$

Although some effort has been made to develop a similar processing based on square-root companding [47, 48, 49, 50, 51, 52, 53, 54], the first option still seems to supply more useful mathematical tools for signal manipulation because of its power function nature (N^x):

$$F(x_1)F(x_2) \equiv F(x_1 + x_2) \tag{1.6}$$

$$\frac{\mathrm{d}F(x)}{\mathrm{d}t} \equiv F(x)\frac{\mathrm{d}x}{\mathrm{d}t} \tag{1.7}$$

Almost all Log companding theories have been implemented using BJT-based circuit techniques for bipolar or BiCMOS technologies like [4, 56, 3, 1, 2, 8, 2, 60, 15, 62, 18, 64, 65, 66, 67, 68, 69, 70, 71, 72, 73, 74, 75, 76, 77, 78, 79, 80, 16, 82, 83, 84, 85, 86, 87, 88, 13, 15, 91, 14, 93, 94, 95, 12, 97]. Unfortunately, CMOS implementations cannot be mapped from these previous bipolar circuits due to important differences at topological, device and also technological levels. In particular:

- Most of the companding bipolar circuit topologies were originally developed to exploit the high-frequency capability of this processing theory and are not low-voltage compatible (e.g. because of the use of cascode structures).
- The MOS device exhibits first-order analog deficiencies with respect to the BJT such as asymmetric I/V curves from input terminals, reduced exponential law DR_I, poor output conductance, physical mismatching and flicker noise.
- Standard CMOS technologies do not usually allow the integration of large value passive devices (i.e. resistors and capacitors), unlike bipolar processes.

Also, the few previous CMOS proposals suffer from a lack of generalization due to local bulks (i.e. separated wells not in compliance with anti-latch-up rules) [98, 10, 100, 101], poor low-voltage operation [102, 103] or redundant circuitry [104, 105].

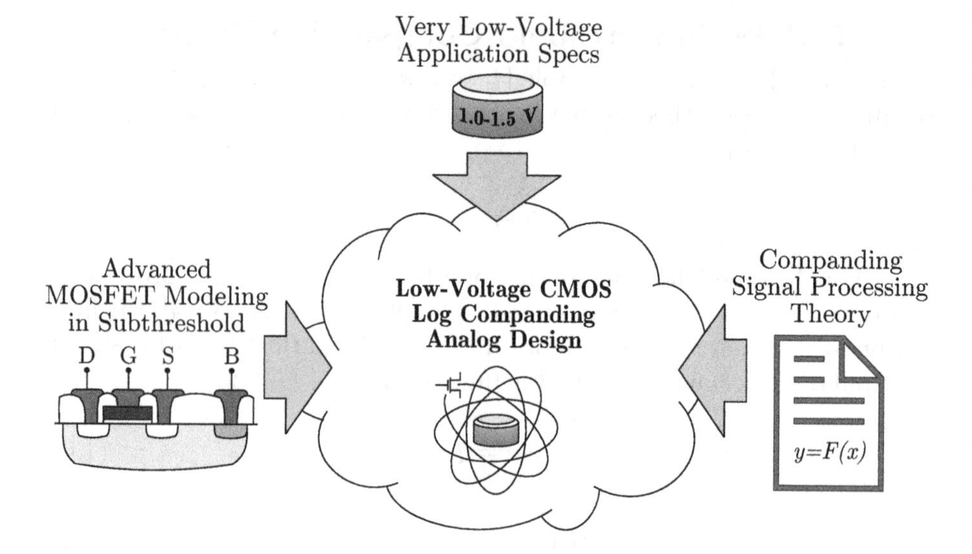

Figure 1.8. Main inputs and outputs of this book.

The aim of this work is the research on novel analog circuit techniques based on the MOSFET operating in subthreshold to exploit the low-voltage capabilities of Log companding signal processing.

The new circuit techniques are based on the MOS transistor operating in its weak inversion region [8] as the basic companding processor. The low current levels available in this region of operation make them suitable for the low-frequency range (up to 100KHz). Hence, target applications for the circuit techniques presented are very low-voltage audio-frequency systems-on-chip, typically battery-powered (e.g. hearing aids). Special attention is paid to the low-voltage performance (down to 1V) of the proposed design techniques.

The main work strategy is shown in Figure 1.8. Since companding processing takes direct advantage of device non-linear characteristics, apart from the Companding Theory itself, a deep knowledge of MOSFET modeling is also needed as previous background.

5. Structure of this Book

In order to define the third input of this work in Figure 1.8, hearing aids were chosen as probably the most restrictive application example in terms of low-voltage specifications. Based on the typical signal processing requirements for these types of systems, the contents of this work have been organized as follows:

Chapter 1 - Introduction. An overview of the low-voltage CMOS analog companding context of the work and its motivations.

Chapter 2 - MOSFET Modeling for Companding. A detailed description of the MOSFET equations used within this publication. Device modeling focuses on Companding Theory implementation, ranging from the analytical model for hand design to its manual extraction procedure.

Chapter 3 - Amplification and AGC. New circuit techniques devoted to signal amplification. Design procedures for gain stages, from general purpose controllable amplifiers to full syllabic AGC systems.

Chapter 4 - Filtering. Novel circuit techniques introduced for signal filtering. Basic building blocks and design methodologies for frequency selective stages, from integrators to arbitrary high-order circuits.

Chapter 5 - PTAT Generation. Circuit techniques proposed for the generation of static PTAT I/V references.

Chapter 6 - Pulse Duration Modulation. Application of the new circuit techniques for computing PDM signals for Class-D output stages.

Chapter 7 - Dynamic Range. An overall and qualitative study of the circuit techniques presented, in terms of signal resolution and distortion.

Chapter 8 - Industrial Application: Hearing Aids. Development of a true 1V CMOS Log-domain analog hearing-aid-on-chip for an industrial customer.

Chapter 9 - Conclusions. General knowledge derived from the results and possible future work in this field.

Appendix A - Simulation and Test. Tips and tricks for numerical simulation and lab setup to measure the performance of the proposed circuits.

Specific background on Companding Theory is supplied at the beginning of each chapter when needed. Apart from the general application example of Chapter 8, illustrative designs with experimental data are given at the end of each chapter as well.

References

[1] The National Technology Roadmap for Semiconductors. Technology Needs. Technical report, Semiconductor Industry Association, 2000. http://www.semichips.org.

[2] H.Baltes and O.Brand. CMOS Integrated Microsystems and Nanosystems. In *Proceedings of the SPIE Conference on Smart Electronics and MEMS*, volume 3673. SPIE, Mar 1999.

[3] A.A.El-Adawy and A.M.Soliman. A Low-Voltage Single Input Class AB Transconductor with Rail-to-Rail Input Range. *IEEE Transactions on Circuits and Systems-II*, 47(2):236–242, Feb 2000.

[4] J.F.Duque-Carrillo, J.L.Alusín, G.Torelli, J.M.Valverde, and M.A.Domínguez. 1-V Rail-to-Rail Operational Amplifiers in Standard CMOS Technology. *IEEE Journal of Solid State Circuits*, 35(1):33–43, Jan 2000.

[5] C.Lin and M.Ismail. Robust Design of LV/LP Low-Distortion CMOS Rail-to-Rail Input Stages. *Journal of Analog Integrated Circuits and Signal Processing, Kluwer Academic Publishers*, 21:153–161, 1999.

[6] Klaas-Jan de Langen and Johan H. Huijsing. Compact Low-Voltage Power-Efficient Operational Amplifier Cells for VLSI. *IEEE Journal of Solid State Circuits*, 33(10):1482–1496, Oct 1998.

[7] G.Ferri, W.Sansen, and V.Peluso. A Low-Voltage Fully-Differential Constant-Gm Rail-to-Rail CMOS Operational Amplifier. *Journal of Analog Integrated Circuits and Signal Processing, Kluwer Academic Publishers*, 16:5–15, 1998.

[8] G.Xu, S.H.K.Embabi, P.Hao, and E.Sánchez-Sinencio. A Low Voltage Fully Differential Nested G_m Capacitance Amplifier: Analysis and Design. In *Proceedings of the International Symposium on Circuits and Systems*, volume 2, pages 606–609. IEEE, 1999.

[9] S.Setty and C.Toumazou. N-Folded Cacode Technique for High Frequency Operation of Low Voltage Opamps. *IEE Electronics Letters*, 32(11):955–956, May 1996.

[10] R.Fried and C.Enz. Nano-Amp, Active Bulk, Weak-Inversion Analog Circuits. In *Proceedings of the Custom Integrated Circuits Conference*, pages 31–34. IEEE, May 1998.

[11] B.J.Blalock, P.E.Allen, and G.A.Rincon-Mora. Designing 1-V Op Amps Using Standard Digital CMOS Technology. *IEEE Transactions on Circuits and Systems-II*, 45(7):769–779, Jul 1998.

[12] G.Dehng, C.Yang, J.Hsu, and S.Liu. A 900-MHz 1-V CMOS Frequency Synthesizer. *IEEE Journal of Solid State Circuits*, 35(8):1211–1214, Aug 2000.

[13] J.Shin, I.Chung, Y.Park, and H.Min. A New Charge Pump Without Degradation in Threshold Voltage Due to Body Effect. *IEEE Journal of Solid State Circuits*, 35(8):1227–1230, Aug 2000.

[14] H.Lin, K.Chang, and S.Wong. Novel High Positive and Negative Pumping. In *Proceedings of the International Symposium on Circuits and Systems*, volume 1, pages 238–241. IEEE, Jun 1999.

[15] R.St.Pierre. Low-Power BiCMOS OpAmp with Integrated Current Mode Charge Pump. In *Proceedings of the European Solid-State Circuits Conference*, pages 70–73. IEEE, Sep 1999.

[16] J.Wu and K.Chang. MOS Charge Pumps for Low-Voltage Operation. *IEEE Journal of Solid State Circuits*, 33(4):592–597, Apr 1998.

[17] T.A.F.Duisters and E.C.Dijkmans. A -90dB THD Rail-to-Rail Input Opamp Using a New Local Charge Pump in CMOS. *IEEE Journal of Solid State Circuits*, 33(7):947–955, Jul 1998.

[18] P.Favrat, P.Deval, and M.J.Declercq. An Improved Voltage Doubler in a Standard CMOS Technology. In *Proceedings of the International Symposium on Circuits and Systems*, volume 1, pages 249–252. IEEE, 1997.

[19] J.F.Dickson. On-Chip High-Voltage Generation MNOS Integrated Circuits Using an Improved Voltage Multiplier Technique. *IEEE Journal of Solid State Circuits*, 11(3):374–378, Jun 1976.

[20] J.Sim, C.Lee, X.Jeong, and H.Park. Adaptive Biasing Folded Cascode CMOS OP Amp with Continuous-Time Push-Pull CMFB Scheme. *IEICE Transactions on Electronics*, E80-C(9):1203–1210, Sep 1997.

[21] M.G.Degraune, J.Rijmenents, E.A.Vittoz, and H.J.de Man. Adaptive Biasing CMOS Amplifiers. *IEEE Journal of Solid State Circuits*, 17(3):522–528, Jun 1982.

[22] R.Klinke, B.J.Hosticka, and H.J.Pfleiderer. A Very-High-Slew-Rate CMOS Operational Amplifier. *IEEE Journal of Solid State Circuits*, 24(3):744–746, Jun 1989.

[23] K.Nagaraj. Slew Rate Enhancement Technique for CMOS Output Buffers. *IEE Electronics Letters*, 25(19):1304–1305, Sep 1989.

[24] E.Vittoz and J.Fellroth. CMOS Analog Integrated Circuits Based on Weak Inversion Operation. *IEEE Journal of Solid State Circuits*, 12(3):224–231, Jun 1977.

[25] R.E.Thomas and A.J.Rosa. *Circuits and Signals: An Introduction to Linear and Interface Circuits.* John Wiley and Sons Inc., 1984.

[26] J.O.Voorman. *Continuous-Time Analog Integrated Circuits.* IEEE Press, 1993.

[27] Y.P.Tsividis. Integrated Continuous-Time Filter Design - An Overview. *IEEE Journal of Solid State Circuits*, 29(3):166–176, Mar 1994.

[28] D.Frey. Future Implications of the Log Domain Paradigm. *IEE Proceedings*, 147(1):65–72, Feb 2000.

[29] J.Mulder, W.A.Serdijn, A.C.van der Woerd, and A.H.M. van Roermund. Dynamic Translinear Circuits: An Overview. *Journal of Analog Integrated Circuits and Signal Processing, Kluwer Academic Publishers*, 22:111–126, 2000.

[30] G.Efthivoulidis and Y.Tsividis. Signal Analysis of Externally Linear Filters. In *Proceedings of the International Symposium on Circuits and Systems*, volume 6, pages 65–68. IEEE, 1999.

[31] Y.P.Tsividis. Externally Linear Integrators. *IEEE Transactions on Circuits and Systems-II*, 45(9):1181–1187, Sep 1998.

[32] Y.Tsividis. Externally Linear, Time-Invariant Systems and Their Application to Companding Signal Processors. *IEEE Transactions on Circuits and Systems-II*, 44(2):65–85, Feb 1997.

[33] C.Toumazou, F.J.Lidgey, and D.G.Haigh. *Analogue IC Design: The Current Mode Approach.* Peter Peregrinus Ltd., 1990.

[34] A.S.Sedra and G.W.Roberts. *Current Conveyor Theory and Practice*, chapter 3, pages 93–126. In C.Toumazou et al. [33], 1990.

[35] Barrie Gilbert. A New Wide-Band Amplifier Technique. *IEEE Journal of Solid State Circuits*, 3(4):353–365, Dec 1968.

[36] N.Krishnapura, Y.Tsividis, and D.R.Frey. Simplified Technique for Syllabic Companding in Log-Domain Filters. *IEE Electronics Letters*, 36(15):1257–1259, Jul 2000.

[37] D.Frey. On Instantaneous vs Syllabic Companding in Log Domain Filters. In *Proceedings of the International Symposium on Circuits and Systems*, pages 101–104, 1997.

[38] Y.Tsividis. Minimizing Power Dissipation in Analogue Signal Processors Through Syllabic Companding. *IEE Electronics Letters*, 35(21):1805–1807, Oct 1999.

[39] J.Mulder, W.A.Serdijn, A.C. van der Woerd, and A.H.M. van Roermund. An Instantaneous and Syllabic Companding Translinear Filter. *IEEE Transactions on Circuits and Systems-II*, 45(2):150–154, Feb 1998.

[40] D.R.Frey and Y.P.Tsividis. Syllabically Companding Log Domain Filter Using Dynamic Biasing. *IEE Electronics Letters*, 33(18):1506–1507, Aug 1997.

[41] C.C.Enz and E.A.Vittoz. CMOS Low-Power Analog Design. In *Emerging Technologies: Designing Low Power Digitals Systems*, pages 79–133. IEEE, 1996.

[42] E.M.Blumenkrantz. The Analog Floating Point Technique. In *Symposium on Low Power Electronics*, pages 72–73. IEEE, 1995.

[43] F.Floru. Attack and Release Time Constants in RMS-Based Feedback Compressors. *Journal of the Audio Engineering Society*, 47(10):788–804, Oct 1999.

[44] J.M.Khoury. On the Design of Constant Settling Time AGC Circuits. *IEEE Transactions on Circuits and Systems-II*, 45(3):283–294, Mar 1998.

[45] B.Gilbert. Translinear Circuits: a Proposed Classification. *IEE Electronics Letters*, 11(1):14–16, Jan 1975.

[46] R.W.Adams. Filtering in the Log-Domain. In *63rd AES Conference*, May 1979.

[47] J.N.Babanezhad and G.Temes. A 20-V Four-Quadrant CMOS Analog Multiplier. *IEEE Journal of Solid State Circuits*, 20(6):1158–1168, Dec 1985.

[48] E.A.M.Klumperink and E.Seevinck. MOS Current Gain Cells with Electronically Variable Gain and Constant Bandwidth. *IEEE Journal of Solid State Circuits*, 24(5):1465–1467, Oct 1989.

[49] E.Seevinck and R.J.Wiegerink. Generalized Translinear Circuit Principle. *IEEE Journal of Solid State Circuits*, 26(8):1098–1102, Aug 1991.

[50] R.J.Weigerink. Computer Aided Analysis and Design of MOS Translineal Circuits Operating in Strong Inversion. *Journal of Analog Integrated Circuits and Signal Processing, Kluwer Academic Publishers*, 9(2):181–187, Mar 1996.

[51] J.Mulder, A.C. van der Woerd, W.A.Serdijn, and A.H.M. van Roermund. Current-Mode Companding $\sqrt{x}$-Domain Integrator. *IEE Electronics Letters*, 32(3):198–199, Feb 1996.

[52] S.Hsiao and C.Wu. A 1.2V CMOS Four Quadrant Analog Multiplier. In *Proceedings of the International Symposium on Circuits and Systems*, volume 1, pages 241–244. IEEE, Jun 1997.

[53] J.Mulder, W.A.Serdijn, A.C. van der Woerd, and A.H.M. van Roermund. A 3.3V Current-Controlled $\sqrt{}$-Domain Oscillator. *Journal of Analog Integrated Circuits and Signal Processing, Kluwer Academic Publishers*, 16:17–28, 1998.

[54] M.Eskiyerli and A.Payne. Square Root Domain Filter Design and Performance. *Journal of Analog Integrated Circuits and Signal Processing, Kluwer Academic Publishers*, 22:231–243, 2000.

[55] D.R.Frey. Log-Domain Filtering: an Approach to Current-Mode Filtering. *IEE Proceedings*, 140(6):406–415, Dec 1993.

[56] B.Gilbert. A Precise Four-Quadrant Multiplier with Subnanosecond Response. *IEEE Journal of Solid State Circuits*, 3(4):365–373, Dec 1968.

[57] Barrie Gilbert. *Current-mode Circuits From a Translinear Viewpoint: A Tutorial*, chapter 2, pages 11–92. In C.Toumazou et al. [33], 1990.

[58] E.Seevinck. Companding Current-Mode Integrator: A New Circuit Principle for Continuous-Time Monolithic Filters. *IEE Electronics Letters*, 26(24):2046–2047, 1990.

[59] A. van Staveren and A.H.M. van Roermund. Low-Voltage Low-Power Controlled Attenuator for Hearing Aids. *IEE Electronics Letters*, 29(15):1355–1356, Jul 1993.

[60] R.Otte and A.H.M. van Roermund. A Low-Voltage, Low-Power Controllable Current Amplifier for Hearing Instruments. In *Proceedings of the European Solid-State Circuits Conference*. IEEE, Sep 1993.

[61] A.C. van der Woerd and W.A.Serdijn. Low-Voltage Low-Power Controllable Preamplifier for Electret Microphones. *IEEE Journal of Solid State Circuits*, 29(9):1052–1055, Oct 1993.

[62] R.Otte and A.H.M. van Roermund. Low-Voltage, Low-Power, Wide-Range Controllable Current Amplifier for Hearing Aids. *IEE Electronics Letters*, 30(3):178–180, Feb 1994.

[63] D.R.Frey. Current Mode Class AB Second Order Filter. *IEE Electronics Letters*, 30(3):205–206, Feb 1994.

[64] D.Perry and G.W.Roberts. Log-Domain Filters Based on LC Ladder Synthesis. In *Proceedings of the International Symposium on Circuits and Systems*, pages 311–314. IEEE, 1995.

[65] J.Mahattanakul, C.Toumazou, and S.Pookaiyaudom. Low-Distortion Current-Mode Companding Integrator Operating at f_T of BJT. *IEE Electronics Letters*, 32(21):2019–2021, Oct 1996.

[66] D.R.Frey. A 3.3 Volt Electronically Tunable Active Filter Usable to Beyond 1GHz. In *Proceedings of the International Symposium on Circuits and Systems*, pages 493–496. IEEE, 1996.

[67] D.R.Frey. Log Filtering Using Gyrators. *IEE Electronics Letters*, 32(1):26–28, Jan 1996.

[68] F.Yang, C.Enz, and G. van Ruymbek. Design of Low-Power and Low-Voltage Log-Domain Filters. In *Proceedings of the International Symposium on Circuits and Systems*. IEEE, 1996.

[69] J.Ngarmnil and C.Toumazou. Micropower Log-Domain Active Inductor. *IEE Electronics Letters*, 32(11):953–955, May 1996.

[70] D.R.Frey and L.Steigerwald. An Adpative Analog Notch Filter Using Log Filtering. In *Proceedings of the International Symposium on Circuits and Systems*. IEEE, 1996.

[71] J.Mahattanakul and C.Toumazou. Instantaneous Companding and Expressing: A Dual Approach to Linear Integrator Synthesis. *IEE Electronics Letters*, 33(1):4–5, Jan 1997.

[72] G. van Ruymbeke, C.C.Enz, F.Krummenacher, and M.Declerq. A BiCMOS Programmable Continuous-Time Filter Using Image-Parameter Method Synthesis and Voltage-Companding Technique. *IEEE Journal of Solid State Circuits*, 32(3):377–387, Mar 1997.

[73] R.Fox, M.Nagarajan, and J.Harris. Practical Design of Single-Ended Log-Domain Filter Circuits. In *Proceedings of the International Symposium on Circuits and Systems*, volume 1, pages 341–344. IEEE, Jun 1997.

[74] J.Mahattanakul and C.Toumazou. DC Stable CCII-Based Instantaneous Companding Integrator. In *Proceedings of the International Symposium on Circuits and Systems*, volume 1, pages 821–824. IEEE, Jun 1997.

[75] J.Mahattanakul and C.Toumazou. A Non-Linear Design Approach for High-Frequency Linear Integrators. In *Proceedings of the International Symposium on Circuits and Systems*, volume 1, pages 485–488. IEEE, Jun 1997.

[76] J.Mahattanakul and C.Toumazou. Modular Log-Domain Filters. *IEE Electronics Letters*, 33(13):1130–1131, Jun 1997.

[77] E.M.Drakakis, A.J.Payne, and C.Toumazou. Log-Domain Filters, Translinear Circuits and the Bernoulli Cell. In *Proceedings of the International Symposium on Circuits and Systems*, volume 1, pages 501–504. IEEE, 1997.

[78] M.El-Gamal and G.W.Roberts. LC Ladder-Based Synthesis of Log-Domain Bandpass Filters. In *Proceedings of the International Symposium on Circuits and Systems*, volume 1, pages 105–108. IEEE, 1997.

[79] M.El-Gamal, V.Leung, and G.W.Roberts. Balanced Log-Domain Filters for VHF Applications. In *Proceedings of the International Symposium on Circuits and Systems*, volume 1, pages 493–496. IEEE, 1997.

[80] D.Frey. State Space Synthesis of Log Domain Filters. In *Proceedings of the International Symposium on Circuits and Systems*, volume 1, pages 481–484. IEEE, 1997.

[81] M.Punzenberger and C.C.Enz. A 1.2-V Low-Power BiCMOS Class AB Log-Domain Filter. *IEEE Journal of Solid State Circuits*, 32(12):1968–1978, Dec 1997.

[82] W.A.Serdijn, M.Broest, J.Mulder, A.C. van der Woerd, and A.H.M. van Roermund. A Low-Voltage Ultra-Low-Power Translinear Integrator for Audio Filter Applications. *IEEE Journal of Solid State Circuits*, 32(4):577–581, Apr 1997.

[83] L.P.L.Dijk, A.C. van der Woerd, and A.H.M.Roermund. An Ultra-Low-Power, Low-Voltage Electronic Audio Delay Line for Use in Hearing Aids. *IEEE Journal of Solid State Circuits*, 32(2):291–294, Feb 1998.

[84] A.Payne, A.Thanachayanont, and C.Papavassilliou. A 150-MHz Translinear Phase-Locked Loop. *IEEE Transactions on Circuits and Systems-II*, 45(9):1220–1230, Sep 1998.

[85] A.Worapishet and C.Toumazou. f_T Integrator - A New Class of Tunable Low-Distortion Instantaneous Companding Integrators for Very High-Frequency Applications. *IEEE Transactions on Circuits and Systems-II*, 45(9):1212–1219, Sep 1998.

[86] M.Punzenberg and C.C.Enz. A Compact Low-Power BiCMOS Log-Domain Filter. *IEEE Journal of Solid State Circuits*, 33(7):1123–1128, Jul 1998.

[87] D.R.Frey. State-Space Synthesis and Analysis of Log-Domain Filters. *IEEE Transactions on Circuits and Systems-II*, 45(9):1205–1211, Sep 1998.

[88] M.N.El-Gamal and G.W.Roberts. Very High-Frequency Log-Domain Band-pass Filters. *IEEE Transactions on Circuits and Systems-II*, 45(9):1188–1198, Sep 1998.

[89] J.Mahattanakul and C.Toumazou. Modular Log-Domain Filters Based upon Linear Gm-C Filter Synthesis. *IEEE Transactions on Circuits and Systems-I*, 46(12):1421–1430, Dec 1999.

[90] P.J.Poort, W.A.Serdijn, J.Mulder, A.C. van der Woerd, and A.H.M. van Roer-mund. A 1-V Class-AB Translinear Integrator for Filter Applications. *Journal of Analog Integrated Circuits and Signal Processing, Kluwer Academic Publishers*, 21:79–90, 1999.

[91] E.M.Drakakis, A.J.Payne, and C.Toumazou. Log-Domain Filtering and the Bernoulli Cell. *IEEE Transactions on Circuits and Systems-I*, 46(5):559–571, May 1999.

[92] D.R.Frey and A.T.Tola. A State-Space Formulation for Externally Linear Class AB Dynamical Circuits. *IEEE Transactions on Circuits and Systems-II*, 46(3):306–314, Mar 1999.

[93] D.A.Panagiotopulos, R.W.Newcomb, and S.K.Singh. A Current-Mode Exponential Amplifier. *IEEE Transactions on Circuits and Systems-II*, 47(6):548–551, Jun 2000.

[94] R.T.Edwards and G.Cauwenberghs. Synthesis of Log-Domain Filters from First-Order Building BLocks. *Journal of Analog Integrated Circuits and Signal Processing, Kluwer Academic Publishers*, 22:177–186, 2000.

[95] A.Worapishet, J.Mahattanakul, and C.Toumazou. A Very High-Frequency Transistor-Only Linear Tunable Companding Current-Mode Integrator. *Journal of Analog Integrated Circuits and Signal Processing, Kluwer Academic Publishers*, 22:187–193, 2000.

[96] A.T.Tola and D.R.Frey. A Study of Different Class AB Log Domain First Order Filters. *Journal of Analog Integrated Circuits and Signal Processing, Kluwer Academic Publishers*, 22:163–176, 2000.

[97] E.M.Drakakis and A.J.Payne. A Bernoulli Cell-Based Investigation of the Non-Linear Dynamics in Log-Domain Structures. *Journal of Analog Integrated Circuits and Signal Processing, Kluwer Academic Publishers*, 22:127–146, 2000.

[98] C.Toumazou, J.Ngarmnil, and T.S.Lande. Micropower Log-Domain Filter for Electronic Cochlea. *IEE Electronics Letters*, 30(22):1839–1841, Oct 1994.

[99] R.Fried, D.Python, and C.C.Enz. Compact Log-Domain Current Mode Integrator with High Transconductance-to-Bias Current Ratio. *IEE Electronics Letters*, 32(11):952–953, May 1996.

[100] D.Masmoudi, W.A.Serdijn, J.Mulder, A.C. van der Woerd, J.Tomas, and J.P.Dom. A New Current-Mode Synthesis Method for Dynamic Translinear Filters and its Applications in Hearing Instruments. *Journal of Analog Integrated Circuits and Signal Processing, Kluwer Academic Publishers*, 22:221–229, 2000.

[101] E.I.El-Masry and J.Wu. Low Voltage Micropower Log-Domain Filters. *Journal of Analog Integrated Circuits and Signal Processing, Kluwer Academic Publishers*, 22:209–220, 2000.

[102] R.M.Fox and M.Nagarajan. Multiple Operating Points in CMOS Log-Domain Filter. *IEEE Transactions on Circuits and Systems-II*, 46(6):705–710, Jun 1999.

[103] D.Frey. C-Log Domain Filters. In *Proceedings of the International Symposium on Circuits and Systems*, volume I, pages 176–179. IEEE, May 2000.

[104] D.Python, M.Punzenberger, and C.Enz. A 1-V CMOS Log-Domain Integrator. In *Proceedings of the International Symposium on Circuits and Systems*, volume II, pages 685–688. IEEE, 1999.

[105] E.Fragnière and E.Vittoz. A Log-Domain CMOS Transcapacitor: Design, Analysis and Applications. *Journal of Analog Integrated Circuits and Signal Processing, Kluwer Academic Publishers*, 22:195–208, 2000.

[106] Y.P.Tsividis. *Operation and Modeling of the MOS Transistor.* Electrical Engineering Series. McGraw-Hill International Editions, 1987.

Chapter 2

MOSFET MODELING FOR COMPANDING

Abstract The contents of this chapter focus on the MOS device background needed to develop the analog circuit techniques presented in this work. Special attention is paid to the analytical large signal I/V model as the main tool for circuit design when implementing the Companding Theory. A specific extraction procedure is also introduced to obtain the required technological parameters either from numerical models or from experimental data.

1. Model Requirements for Analytical Design

Historically, numerical simulation has always been a bottleneck in the design of integrated circuits since it requires a proper balance between accuracy and computation time. A good example of this compromise is today's most used CAD standard for microelectronics called Simulation Program with Integrated Circuit Emphasis (SPICE) [1].

However, from the analog point of view, not only high numerical accuracy is needed for the analysis, but also analytical device models for the synthesis process. If the designer has no access to a simple set of device equations to operate mathematically with them in a user-friendly way, analog design becomes a kind of trial-and-error procedure between handwork and numerical simulation as pointed out in [2], which is quite time-consuming and does not return any optimum circuit topology.

This necessity is of particular importance for the analog circuit techniques developed in this work since the Companding Theory directly exploits the device-level characteristics to perform its signal processing. Hence, an accurate but easy-to-use model has to be chosen for the MOS transistor. Standard SPICE models for industrial CMOS technologies such as the public domain BSIM3 [1, 2] exhibit good enough accuracy for any simulation tasks of this work. However, this accuracy and computation efficiency have been increased by using a larger number of fitting parameters, thus making the equations more complex.

As a result, the first task of this work was to set up a simple but accurate analytical model for the bulk enhancement metal-oxide-Silicon field effect transistor (MOSFET), always keeping in mind the analog companding design point of view. Special attention must to be paid to the large signal I/V equations since their non-linearities are the basis for the CMOS analog circuit techniques presented. The main features of the ideal device model are as follows:

Local Bulk terminal as the reference potential for the MOSFET symbol, instead of the classical common-source structure (no V_{GS} nomenclature!). This scheme is suitable for the anti-latch-up rules of CMOS processes and gives more powerful equations to design circuits with MOS devices exhibiting non-grounded drains or sources.

Explicit expressions, making the body effect ($V_{SB} \neq 0$) useful for design purposes, for example.

Bidirectionality from drain-to-source in accordance with the particular physical symmetry of the MOS transistor from these two terminals.

Single expressions for all regions of operation. The asymptotic behavior of this model in each particular region should match the limited-range equations in classical models.

Continuous first derivatives around all the boundaries of the different regions of operation, which prevents discontinuity in all small signal parameters.

In this sense, all equations in this chapter refer to the device nomenclature of Figure 2.1, where I_D, V_{DB}, V_{GB} and V_{SB} stand for drain current, drain-to-bulk, gate-to-bulk and source-to-bulk voltages, respectively.

2. Large Signal Equations

The first step toward a suitable MOSFET modeling for companding design is related to the non-linear behaviour of such a transistor, which allows direct implementation at the device level of the inner non-linear processing within the Log domain.

2.1 DC Drain Current

For the Companding Theory, the most interesting characteristic of the MOS device is the static non-linear drain current as a function of the

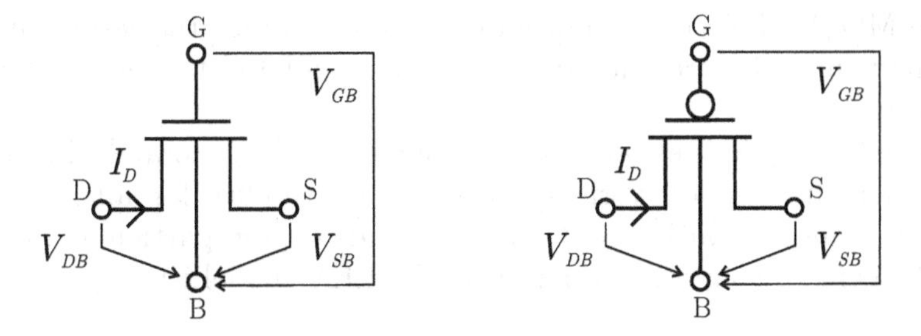

Figure 2.1. Basic nomenclature for NMOS (left) and PMOS (right) devices.

terminal voltages. The following expressions are a simplification of the full EKV model [5, 6] which serves to reduce the number of technological parameters needed and to facilitate its usage during hand design (see [7] for more in-depth knowledge on the physical basis of this model).

In order to eliminate redundant formulae, only the NMOS-type equations will be described, although the dual PMOS case is directly obtained supposing opposite sign in the threshold voltage and the following symmetry:

$$I_D\left(V_{GB}, V_{SB}, V_{DB}\right)|_{PMOS} \equiv -I_D\left(-V_{GB}, -V_{SB}, -V_{DB}\right)|_{NMOS} \quad (2.1)$$

The total I_D can be understood as the sum of two opposite components, forward (I_{DF}) and reverse (I_{DR}), which only depend on the (V_{GB},V_{SB}) and (V_{GB},V_{DB}) biasing, respectively, according to:

$$I_D = I_{DF} - I_{DR} = I_S\left[F\left(\frac{V_P - V_{SB}}{U_t}\right) - F\left(\frac{V_P - V_{DB}}{U_t}\right)\right] \quad (2.2)$$

$$V_P = \frac{V_{GB} - V_{TO}}{n} \quad (2.3)$$

$$I_S = 2n\beta {U_t}^2 \quad (2.4)$$

where V_P and V_{TO} stand for the well-known pinch-off and threshold Voltages, while I_S, n, β and U_t correspond to the specific current, subthreshold slope, current factor (for a channel aspect ratio W/L) and thermal potential (about 25mV at room temperature) respectively.

It is important to note the equivalence between the above functions and the general companding formulae introduced in Chapter 1. Thus, the idea of implementing the theoretical companding function F through the non-linear shape of the device I/V curve can be seen more clearly.

In the MOSFET case, the normalizing constants for the y-domain and x-domain in (1.3) are the specific current and the thermal potential, respectively.

While the explicitness and bidirectionality of the general drain current expression are already verified by (2.2), its validity for all regions of operation and derivative continuity depends on the particular function F. The general solution proposed by the EKV model is:

$$y = F(x) = \ln^2\left(1 + e^{\frac{x}{2}}\right) = \begin{cases} e^x & x \ll 0 \\ \left(\frac{x}{2}\right)^2 & x \gg 0 \end{cases} \tag{2.5}$$

Hence, the final $I_D\left(V_{GB}, V_{SB}, V_{DB}\right)$ can be rewritten as:

$$I_D = I_S\left[\ln^2\left(1 + e^{\frac{V_{GB}-V_{TO}-nV_{SB}}{2nU_t}}\right) - \ln^2\left(1 + e^{\frac{V_{GB}-V_{TO}-nV_{DB}}{2nU_t}}\right)\right] \tag{2.6}$$

This analytical and yet simple expression gives the analog designer a powerful mathematical tool for the synthesis process. For example, applying the asymptotes of F from (2.5) to (2.6), it is easy now to obtain the drain current expression for each region of operation just by hand manipulation. The exhaustive results are listed in Table 2.1 for weak, moderate and strong inversion as well as for conduction, forward and reverse saturation. For our purposes, the role of the smoothing function F is more important in the boundary between weak and strong inversion regions in saturation. Figure 2.2 shows this fact graphically, where moderate inversion can be identified as the region where no single asymptote approximation is dominant. Instead of using the voltage boundaries of Table 2.1, a current reference is preferred to identify the above inversion regions. In this sense, the inversion coefficient is defined as:

$$IC = \frac{I_D}{I_S} \tag{2.7}$$

Now, the IC parameter lets the designer locate and compare the regions of operation for MOS transistors through a logarithmic scale, without being concerned about the particular values of the terminal voltages. For example, the moderate inversion region can be approximately identified with one decade around $IC = 0\text{dB}$, as depicted in Figure 2.2.

Second-order effects must be considered when using small MOSFET geometries [8, 1] such as velocity saturation, lateral bird-peaks or diffusion charge sharing. However, the channel length modulation (CLM) [8]

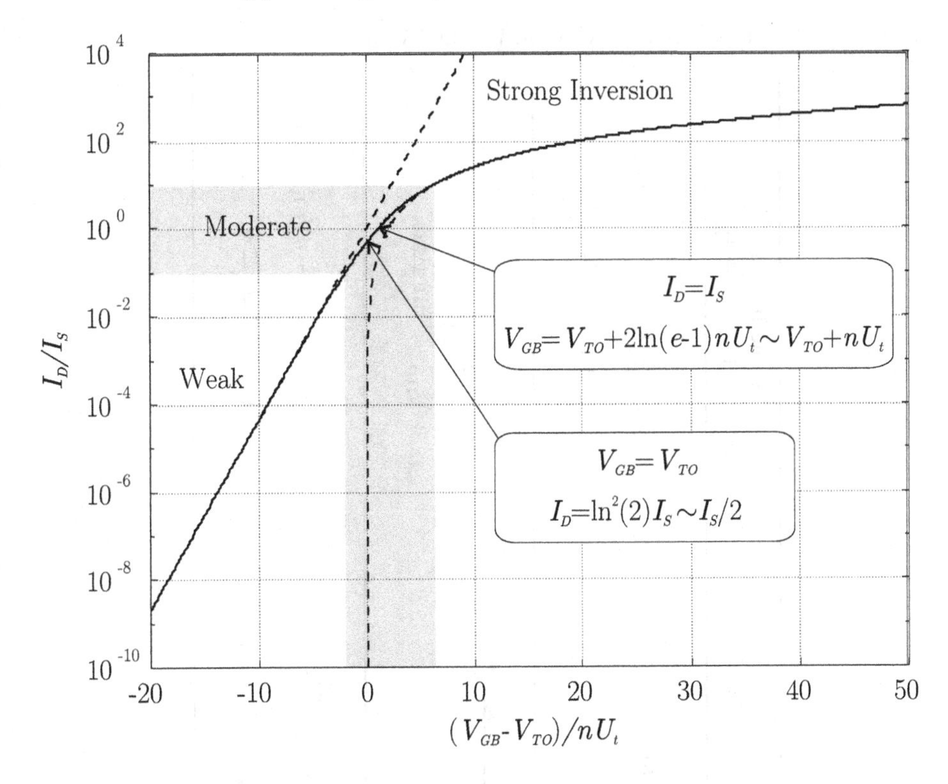

Figure 2.2. Normalized drain current versus pinch-off voltage in forward saturation and $V_{SB} = 0\text{V}$. The dashed lines indicate the asymptotic approximations of Table 2.1.

is perhaps the most important effect in analog environments, since it tends to lower the output device impedance in saturation according to:

$$I_{Deff} = \begin{cases} I_D\left(1 + \lambda V_{DB}\right) & \text{forward sat.} \\ I_D\left(1 + \lambda V_{SB}\right) & \text{reverse sat.} \end{cases} \tag{2.8}$$

where λ stands for the CLM parameter and is inversely proportional to the channel length L.

2.2 Quasi-Static Capacitances

The static I/V description of the MOS device presented in the previous subsection does not include by definition any time dependence, so some additional information is needed in order to predict its transient behavior. Such dynamic large signal modeling is performed through the non-linear voltage-dependent intrinsic MOS capacitances. Since the final target of the application is located at the low-frequency range (up to

Table 2.1. DC large signal I_D vs V_{GB}, V_{SB} and V_{DB}.

	Weak Inversion V_{SB} and $V_{DB} \gg \frac{V_{GB}-V_{TO}}{n}$	Moderate Inversion otherwise	Strong Inversion V_{SB} or $V_{DB} \ll \frac{V_{GB}-V_{TO}}{n}$
Conduction	$I_S e^{\frac{V_{GB}-V_{TO}}{nU_t}}(e^{-\frac{V_{SB}}{U_t}} - e^{-\frac{V_{DB}}{U_t}})$ $\lvert V_{DB} - V_{SB}\rvert \ll U_t$	expression (2.6) $I_{DF} \sim I_{DR}$	$\beta[(V_{GB} - V_{TO}) - \frac{n}{2}(V_{DB} + V_{SB})](V_{DB} - V_{SB})$ V_{SB} and $V_{DB} \ll \frac{V_{GB}-V_{TO}}{n}$
Forward Sat.	$I_S e^{\frac{V_{GB}-V_{TO}}{nU_t}} e^{-\frac{V_{SB}}{U_t}}$ $(V_{DB} - V_{SB}) \gg U_t$	$I_S \ln^2(1 + e^{\frac{V_{GB}-V_{TO}-nV_{SB}}{2nU_t}})$ $I_{DF} \gg I_{DR}$	$\frac{\beta}{2n}(V_{GB} - V_{TO} - nV_{SB})^2$ $V_{DB} \gg \frac{V_{GB}-V_{TO}}{n}$
Reverse Sat.	$-I_S e^{\frac{V_{GB}-V_{TO}}{nU_t}} e^{-\frac{V_{DB}}{U_t}}$ $(V_{SB} - V_{DB}) \gg U_t$	$-I_S \ln^2(1 + e^{\frac{V_{GB}-V_{TO}-nV_{DB}}{2nU_t}})$ $I_{DF} \ll I_{DR}$	$-\frac{\beta}{2n}(V_{GB} - V_{TO} - nV_{DB})^2$ $V_{SB} \gg \frac{V_{GB}-V_{TO}}{n}$

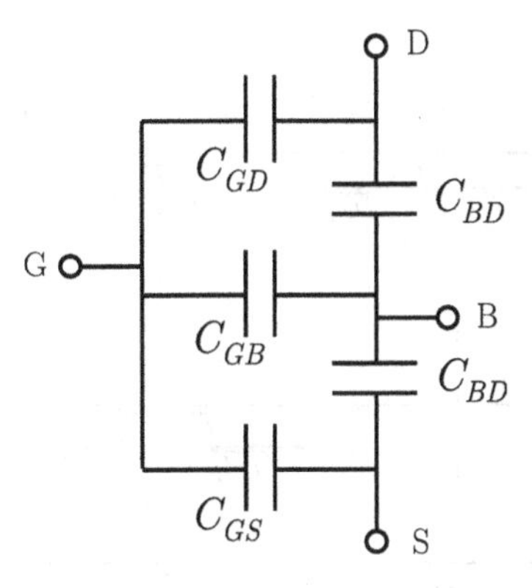

Figure 2.3. Simplified transcapacitance MOS model.

100KHz), redistribution time constants of non-quasi-static models [8, 9] can be neglected here. Furthermore, the MOS capacitance model is not used in this work to study high-frequency parasitic elements, but to evaluate signal distortion in all-MOS companding proposals.

The interesting subset of MOSFET transcapacitances for our purposes is depicted in Figure 2.3 and may be defined as:

$$C_{GS} \doteq -\left.\frac{\partial Q_G}{\partial V_{SB}}\right|_{V_{GB0},V_{DB0}} \qquad C_{GD} \doteq -\left.\frac{\partial Q_G}{\partial V_{DB}}\right|_{V_{GB0},V_{SB0}} \qquad C_{GB} \doteq -\left.\frac{\partial Q_G}{\partial V_{GB}}\right|_{V_{GS0},V_{GD0}} \tag{2.9}$$

$$C_{BS} \doteq -\left.\frac{\partial Q_B}{\partial V_{SB}}\right|_{V_{GB0},V_{DB0}} \qquad C_{BD} \doteq -\left.\frac{\partial Q_B}{\partial V_{DB}}\right|_{V_{GB0},V_{SB0}} \tag{2.10}$$

where Q stands for charge. In general, all the above elements are non-reciprocal (i.e. $C_{MN} \neq C_{NM}$) due to the MOS device asymmetries (e.g. gate versus source terminals). Since bulk-charge related C_{BS} and C_{BD} can be obtained from:

$$C_{BD} \equiv (n-1)C_{GD} \qquad C_{BS} \equiv (n-1)C_{GS} \tag{2.11}$$

only the three transcapacitances controlling the charge in the isolated gate are developed. The following expressions for all regions of operation are extracted from the general EKV model:

$$C_{GD} = C_{ox} \left[\frac{1}{\frac{2}{3}(1 - \frac{IC_F}{(\sqrt{IC_F}+\sqrt{IC_R})^2})} + \frac{1}{\sqrt{IC_R}\left(1 - e^{-\sqrt{IC_R}}\right)} \right]^{-1} \quad (2.12)$$

$$C_{GB} = C_{ox}\frac{n-1}{n}\left[1 - \left[\frac{1}{\frac{2}{3}(1 + 2\frac{\sqrt{IC_F IC_R}}{(\sqrt{IC_F}+\sqrt{IC_R})^2})} + \frac{1}{\sqrt{IC_F}(1 - e^{-\sqrt{IC_F}}) + \sqrt{IC_R}(1 - e^{-\sqrt{IC_R}})}\right]^{-1}\right] \quad (2.13)$$

$$C_{GS} = C_{ox} \left[\frac{1}{\frac{2}{3}(1 - \frac{IC_R}{(\sqrt{IC_F}+\sqrt{IC_R})^2})} + \frac{1}{\sqrt{IC_F}\left(1 - e^{-\sqrt{IC_F}}\right)} \right]^{-1} \quad (2.14)$$

where C_{ox} stands for the nominal oxide capacitance, while IC_F and IC_R symbolize the forward and reverse inversion coefficients defined according to (4.57):

$$IC_F \doteq \frac{I_F}{I_S} \qquad IC_R \doteq \frac{I_R}{I_S} \quad (2.15)$$

The asymptotic values for each region of operation can be again derived just by hand manipulation. The exhaustive results are listed in Table 2.2 and Figure 2.4.

3. Small signal Parameters

The following information is intended to be mainly used in this work for the design of the auxiliary control circuitry around the basic CMOS companding topologies. The main tools for studying its loop gain, stability and other frequency domain specifications are the set of transconductance parameters shown in Figure 2.5. Their definition comes from the linear incremental model:

$$g_{mg} \doteq \left.\frac{\partial I_D}{\partial V_{GB}}\right|_{V_{SB0},V_{DB0}} \qquad g_{ms} \doteq -\left.\frac{\partial I_D}{\partial V_{SB}}\right|_{V_{GB0},V_{DB0}} \qquad g_{md} \doteq \left.\frac{\partial I_D}{\partial V_{DB}}\right|_{V_{GB0},V_{SB0}} \quad (2.16)$$

Table 2.2. Asymptotes of the normalized MOS transcapacitances ($\frac{C_{GD}}{C_{ox}}, \frac{C_{GB}}{C_{ox}}, \frac{C_{GS}}{C_{ox}}$) versus the same region boundaries as in Table 2.1. In conduction cases, $IC_F = IC_R \doteq IC_{FR}$ has been assumed for simplification.

	Weak Inversion	Strong Inversion
Conduction	$(IC_{FR}, \frac{n-1}{n}, IC_{FR})$	$(\frac{1}{2}, 0, \frac{1}{2})$
Forward Sat.	$(0, \frac{n-1}{n}, IC_F)$	$(0, \frac{n-1}{3n}, \frac{2}{3})$
Reverse Sat.	$(IC_R, \frac{n-1}{n}, 0)$	$(\frac{2}{3}, \frac{n-1}{3n}, 0)$

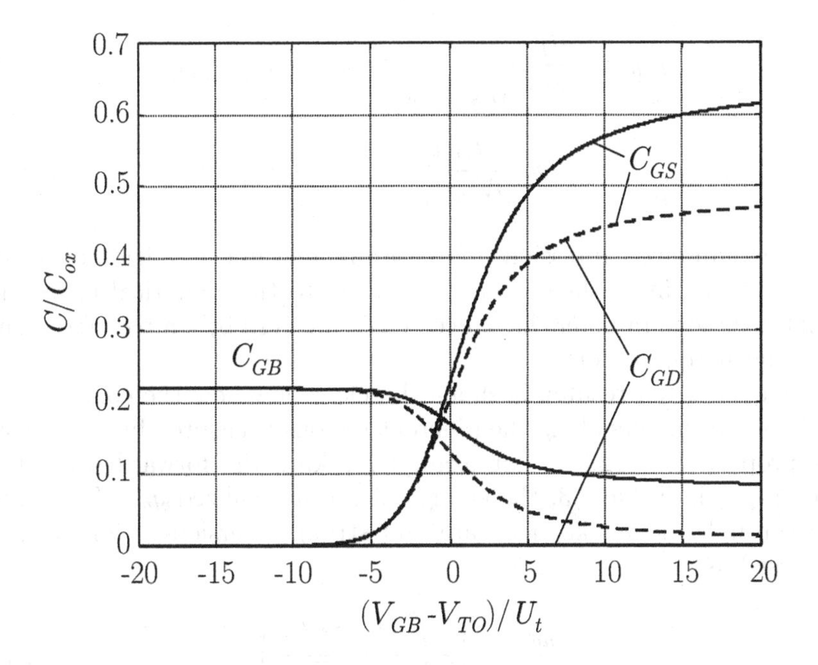

Figure 2.4. Normalized MOS transcapacitances in conduction (dashed) and forward saturation (solid) for $n = 1.3$.

The equivalence of the above parameters with the classical set of transconductances referred to the source instead of to the bulk is as follows:

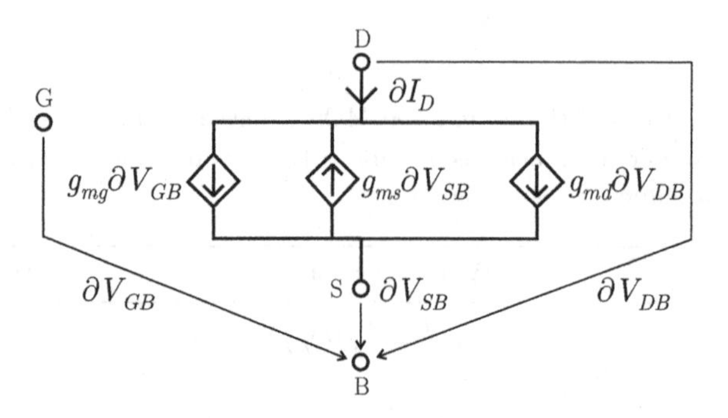

Figure 2.5. DC small signal equivalent circuit of the CMOS transistor.

$$
\begin{aligned}
g_m &\doteq \left.\frac{\partial I_D}{\partial V_{GS}}\right|_{V_{BSO},V_{DSO}} \equiv g_{mg} \\
g_{mb} &\doteq \left.\frac{\partial I_D}{\partial V_{BS}}\right|_{V_{GSO},V_{DSO}} \equiv g_{ms} - g_{mg} - g_{md} \qquad (2.17) \\
g_{ds} &\doteq \left.\frac{\partial I_D}{\partial V_{DS}}\right|_{V_{GSO},V_{BSO}} \equiv g_{md}
\end{aligned}
$$

The exhaustive expressions of all parameters in each region of operation are listed in Table 2.3. Again, thanks to the analytical properties of (2.6), these parameters have been easily derived by simple mathematical hand manipulation.

From the given results, it should be clear now how the continuity of all these parameters along the different regions is ensured by the derivative properties of (2.5). For example, taking the forward saturation case of g_{mg} in Table 2.3, the change of the normalized g_{mg}/I_D parameter through the weak, moderate and strong inversion regions can be expressed as:

$$
\frac{g_{mg}}{I_D} = \frac{1}{nU_t}\left(\frac{1-e^{-\sqrt{IC}}}{\sqrt{IC}}\right) \qquad (2.18)
$$

The graphical representation of this expression in Figure 2.6 does not suffer from the typical discontinuity in the moderate region exhibited by simpler MOSFET models [10].

Table 2.3. Small signal parameter g_{mg} (upper), g_{ms} (middle) and g_{md} (lower) versus the same region boundaries as in Table 2.1. Expressions in brackets take into account the CLM effect.

	Weak Inversion	Moderate Inversion	Strong Inversion
Conduction	$\frac{I_D}{nU_t}$	$\sqrt{\frac{2\beta}{n}}[\sqrt{I_{DF}}(1-e^{-\sqrt{IC_F}}) -\sqrt{I_{DR}}(1-e^{-\sqrt{IC_R}})]$	$\beta(V_{DB}-V_{SB})$
Forward Sat.	$\frac{I_D}{nU_t}$	$\sqrt{\frac{2\beta I_D}{n}}(1-e^{-\sqrt{IC}})$	$\sqrt{\frac{2\beta I_D}{n}}$
Reverse Sat.	$\frac{I_D}{nU_t}$	$-\sqrt{\frac{2\beta\vert I_D\vert}{n}}(1-e^{-\sqrt{IC}})$	$-\sqrt{\frac{2\beta\vert I_D\vert}{n}}$

	Weak Inversion	Moderate Inversion	Strong Inversion
Conduction	$\frac{I_{DF}}{U_t}$	$\sqrt{2n\beta I_{DF}}[1-e^{-\sqrt{IC_F}}]$	$\sqrt{2n\beta I_{DF}}$
Forward Sat.	$\frac{I_D}{U_t}$	$\sqrt{2n\beta I_D}[1-e^{-\sqrt{IC}}]$	$\sqrt{2n\beta I_D}$
Reverse Sat.	$0 \quad (\lambda I_D)$	$0 \quad (\lambda I_D)$	$0 \quad (\lambda I_D)$

	Weak Inversion	Moderate Inversion	Strong Inversion
Conduction	$\frac{I_{DR}}{U_t}$	$\sqrt{2n\beta I_{DR}}[1-e^{-\sqrt{IC_R}}]$	$\sqrt{2n\beta I_{DR}}$
Forward Sat.	$0 \quad (\lambda I_D)$	$0 \quad (\lambda I_D)$	$0 \quad (\lambda I_D)$
Reverse Sat.	$-\frac{I_D}{U_t}$	$-\sqrt{2n\beta\vert I_D\vert}[1-e^{-\sqrt{IC}}]$	$-\sqrt{2n\beta\vert I_D\vert}$

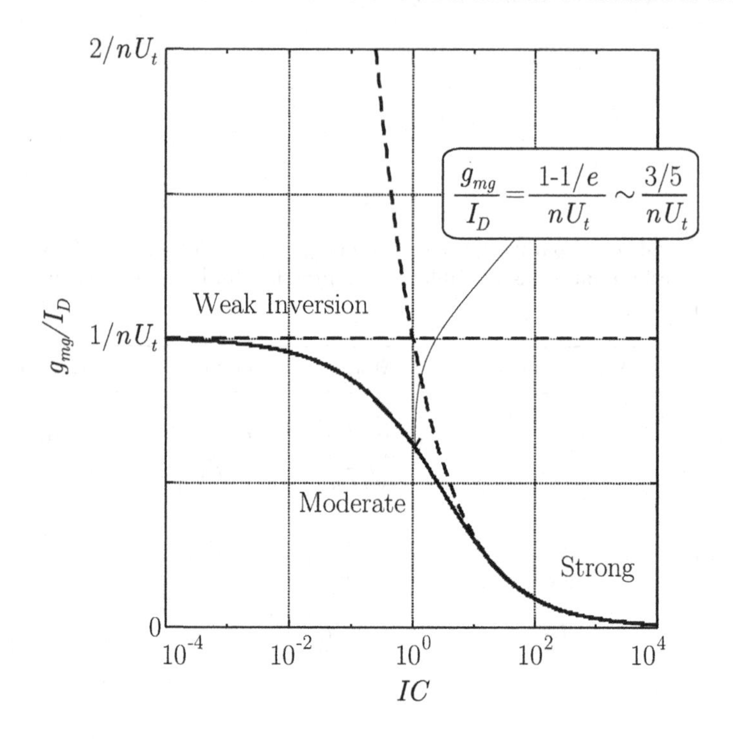

Figure 2.6. Normalized g_{mg} versus inversion coefficient in forward saturation. Dashed lines indicate the asymptotic approximations of Table 2.1

4. Noise Equations

A small signal and spectral model is chosen for the intrinsic noisy phenomena of the MOS transistor. Due to their random nature, the equivalent noise sources are described in terms of power spectral densities (PSD). The main expressions are again derived from the EKV model, although they have been rewritten here as equivalent parallel current sources between drain and source terminals. This nomenclature facilitates dynamic range and signal-to-noise studies in the I-domain for the circuit techniques developed.

Two different types of noise are considered for the MOS device [11]: thermal, caused by the random movement of free electrons in a resistive material, and flicker, originated by fluctuations in the conductivity of imperfect junctions, the $Si\text{-}SiO_2$ interface in the MOSFET case.

The thermal component exhibits a flat or white PSD proportional to the total charge stored in the channel:

$$\frac{\mathrm{d}{I_{DNth}}^2}{\mathrm{d}f} = 4KTg_{ms}K_{th} \tag{2.19}$$

Table 2.4. thermal factor (K_{th}) versus region of operation.

	Weak Inversion	Moderate Inversion	Strong Inversion
Conduction	1	1	1
Forward Sat.	$\frac{1}{2}$	$\frac{3+4IC}{6+6IC}$	$\frac{2}{3}$

$$K_{th} \doteq \frac{1}{1+IC_F}\left(\frac{1+\frac{IC_R}{IC_F}}{2}+\frac{2}{3}IC_F\frac{1+\sqrt{\frac{IC_R}{IC_F}}+\frac{IC_R}{IC_F}}{1+\sqrt{\frac{IC_R}{IC_F}}}\right) \tag{2.20}$$

is the thermal factor, while IC_F and IC_R stand for the forward and reverse inversion coefficients, respectively. The equivalent noise transconductance is interpolated through the different regions of operation via K_{th} as listed in Table (2.4). Reverse saturation has been omitted due to its full duality to the forward case.

On the other hand, flicker noise exhibits a $1/f$ or pink PSD in terms of equivalent gate voltage:

$$\frac{\mathrm{d}V_{GBNfk}{}^2}{\mathrm{d}f} = \frac{K_{fk}}{WL}\frac{1}{f} \tag{2.21}$$

where K_{fk} stands for the flicker factor, with a strong technological dependence (up to 2 orders of magnitude between complementary devices $K_{fkNMOS} \sim 10^2 K_{fkPMOS}$), but very little sensitivity with respect to the drain current. Finally, the total noise drain current can be computed as the root-sum-square of the thermal and flicker components, provided these two phenomena are not correlated:

$$\frac{\mathrm{d}I_{DN}{}^2}{\mathrm{d}f} = \frac{\mathrm{d}I_{DNth}{}^2}{\mathrm{d}f} + \frac{\mathrm{d}I_{DNfk}{}^2}{\mathrm{d}f} = 4KTg_{ms}K_{th} + g_{mg}{}^2\frac{K_{fk}}{WL}\frac{1}{f} \tag{2.22}$$

The dominant range for each noise component is of particular interest when compared to the signal bandwidth, since it can help to select between different circuit strategies. In this sense, the noise corner frequency is defined here as the boundary between thermal and flicker spectral regions (i.e. $\mathrm{d}I_{DNth}{}^2/\mathrm{d}f \equiv \mathrm{d}I_{DNfk}{}^2/\mathrm{d}f$):

$$f_{CN} \doteq \frac{g_{mg}{}^2}{g_{ms}}\frac{K_{fk}}{K_{th}}\frac{1}{4KTWL} \tag{2.23}$$

5. Technology Mismatching Model

Although not usually included in standard simulation models, real integrated MOSFETs manifest non-deterministic variations of their electrical properties. These time-invariant deviations are caused by physical mismatching originated mainly during the manufacturing process. While these effects may not be critical in digital designs, they play an important role in the practical dimensioning of MOS devices for analog circuit techniques based on matching ratios, like the case of this work. As a result, a stochastic model needs to be added to reflect this random behavior.

Only transistor-level technology mismatching will be considered here, that is, differences between two identically sized MOS transistors of the same integrated circuit. Process spread at both batch and run levels are usually covered by the corner models (e.g. worst speed and worst power).

For a general electrical parameter (P), two types of variations (ΔP) are observed:

Local variations with correlation distances comparable to device dimensions. Commonly caused by non-uniform distributions of implanted, diffused or substrate ions, local mobility fluctuations, granular oxide and trapped charges in oxide. They can be fitted to a spatial zero-mean normal distribution.

Global variations when correlation distances are larger than practical device geometries. In this case, the origin can be found in the wafer fabrication and oxidation process. Normal probabilistic distributions are also used to model these type of variations.

Under the assumption of non-correlation, the total deviation of ΔP is the root-sum-square of all contributions. A well accepted model for CMOS technologies [12, 13] is formulated as:

$$\sigma^2(\Delta P) = \frac{{A_P}^2}{WL} + {B_P}^2 D \simeq \frac{{A_P}^2}{WL} \tag{2.24}$$

where A_P and B_P stand for the mismatching parameters of local and global variations, respectively. While the first component depends only on the device area (WL), the latter is related to the distance (D) between the two matched MOS transistors. The above equation does not take into account other important aspects of the layout such as device perimeters, substrate orientation, surrounding layers and thermal gradients. Hence, in order to avoid further geometrical variables, the practical recommendations compiled in Table 2.5 are supposed to be used in the physical design of the analog circuit. In fact, the contributions of global variations can be strongly reduced by minimizing the distance D down to the lithography limits specified in the design rules of the CMOS process. As a result, the total deviation of ΔP can be simplified according to (2.24).

The electrical parameters usually modeled are β and V_{TO}. Also in the context of this work, the mismatch of the drain current is of particular importance when designing the final dimensioning of the MOS devices. The total ΔI_D deviation can be obtained supposing non-correlation between these two electrical parameters:

$$\sigma^2(\Delta I_D) = \sigma^2(\Delta\beta) + {g_{mg}}^2\sigma^2(\Delta V_{TO}) \tag{2.25}$$

When speaking in terms of resolution in the I-domain, a more useful figure is the relative mismatching $\frac{\Delta I_D}{I_D}$ rather than the absolute values. Furthermore, this type of study is commonly interesting when the MOSFET is operating in saturation with high output impedance, hence behaving as a controlled current source. In this case, (2.25) can be rewritten as:

$$\left(\frac{\sigma(\Delta I_D)}{I_D}\right)^2 = \frac{1}{WL}\left[\left(\frac{A_\beta}{\beta}\right)^2 + \left(\frac{1-e^{-\sqrt{IC}}}{\sqrt{IC}}\right)^2\left(\frac{A_{VTO}}{nU_t}\right)^2\right] \tag{2.26}$$

Due to the non-linear law of the drain current versus the threshold voltage, such a current mismatching changes along the saturation regions of operation. Practical values of β and V_{TO} deviations are collected in Figure 2.8 and Figure 2.9 from experimental reports [12, 13, 14, 15, 16, 17, 18, 19, 20, 21, 22]. Based on these data, a comparison between both terms of (2.26) returns a V_{TO} dominance extending from deep weak inversion regions to medium strong inversion regions.

In this sense, a knee inversion coefficient (IC_{knee}) is defined as the boundary between V_{TO} and β dominant mismatching regions in strong inversion:

$$IC_{knee} \simeq \left(\frac{A_{VTO}/nU_t}{A_\beta/\beta}\right)^2 \tag{2.27}$$

An example of a typical behaviour is plotted in Figure 2.7. For a typical 1μm CMOS technology, so that $A_\beta/\beta = 2.5\%\mu$m and $A_{VTO} =$ 15mVμm, the resulting $IC_{knee} \simeq 256$ is locates this boundary at more than two decades over I_S.

Hence, considering only V_{TO} deviations, the asymptotic expressions for each region of saturated inversion are as follows:

Layout Rule	Bad	Good
Unitary Elements	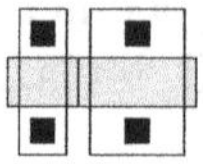	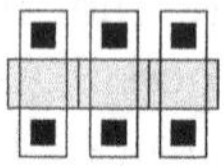
Large Area	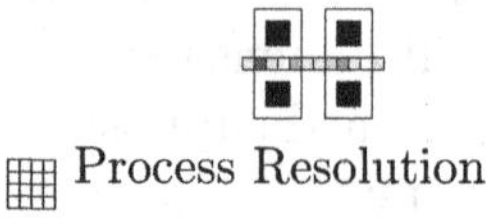 Process Resolution	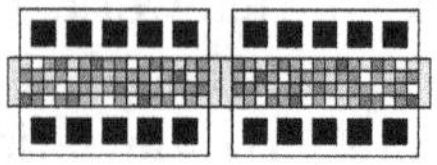
Same Orientation	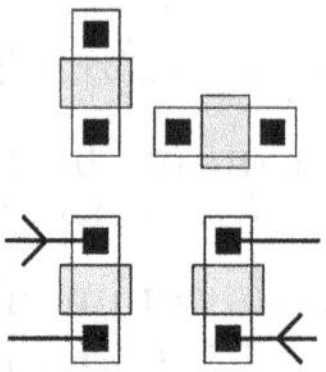	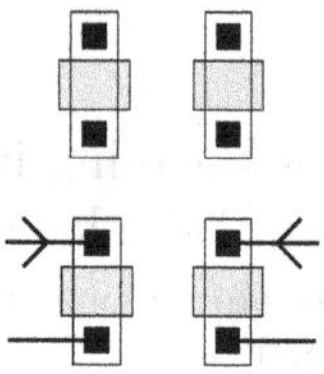
Minimum Distance		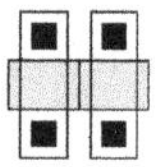
Same Surround	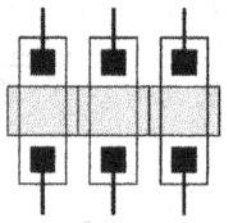	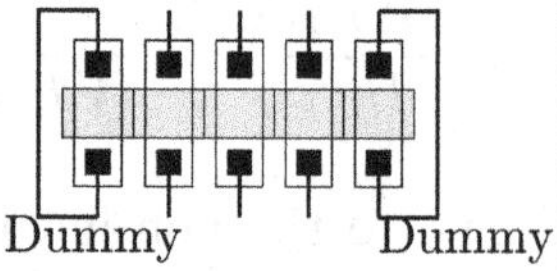 Dummy Dummy
Same Symmetry	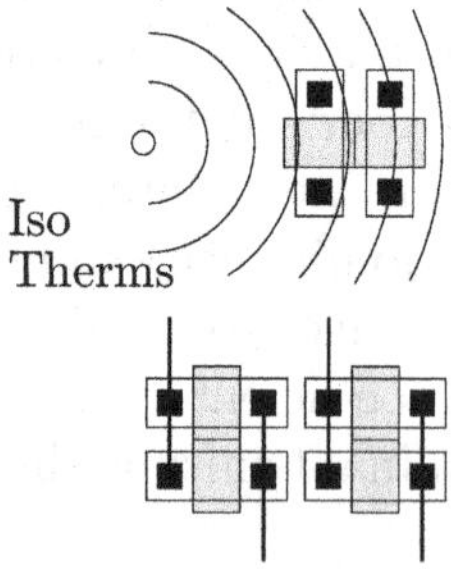 Iso Therms	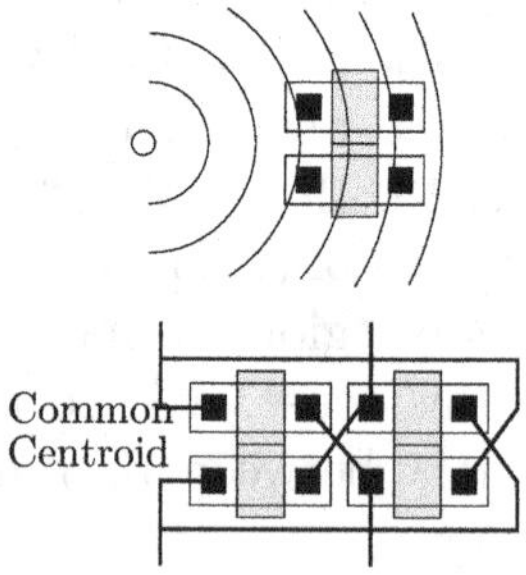 Common Centroid
Examples $(W/L)\gg1$ $(W/L)\ll1$		Interleaved

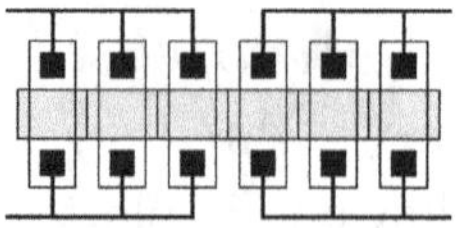

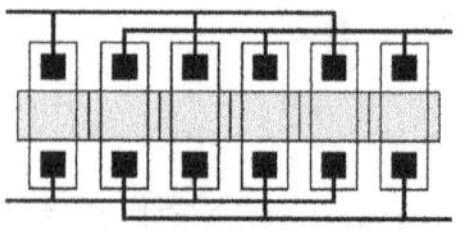

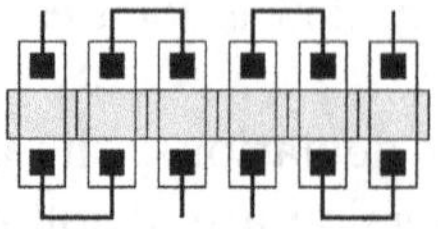

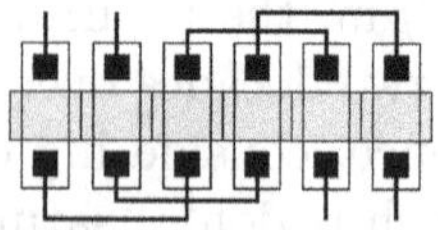

Table 2.5. General layout recommendations for CMOS device matching.

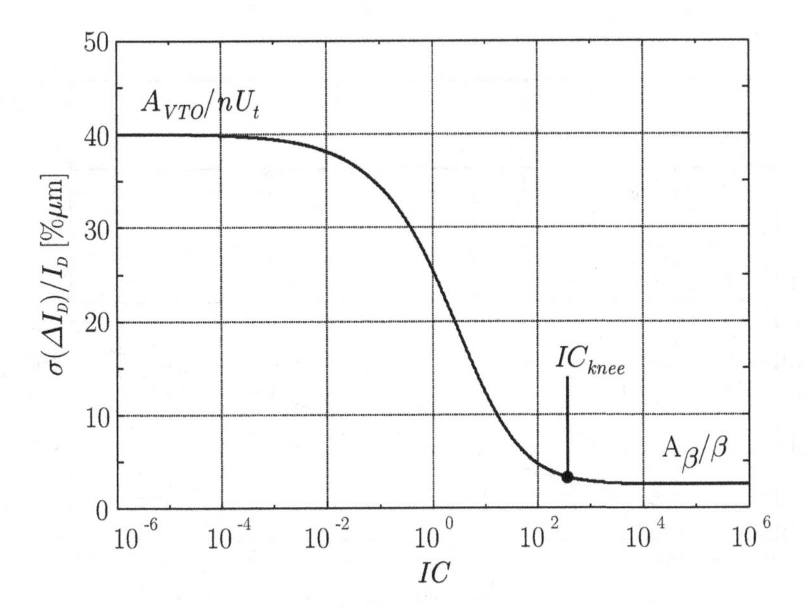

Figure 2.7. Relative drain current deviations versus inversion coefficient for a typical 1μm CMOS process of Figure 2.8 and $n = 1.5$ at room temperature.

$$\frac{\sigma(\Delta I_D)}{I_D} \simeq \begin{cases} \frac{1}{\sqrt{WL}} \frac{A_{VTO}}{nU_t} & IC \ll 1 \\ \frac{1}{L} \frac{A_{VTO}}{\sqrt{nI_D/2\beta}} & 1 \ll IC \ll IC_{knee} \end{cases} \tag{2.28}$$

The maximum and constant relative deviation is reached at weak inversion due to the factor $e^{\frac{\Delta V_{TO}}{nU_t}}$ in Table 2.1. On the other hand, the sensitivity versus V_{TO} tends to decrease for larger voltage overdrives. Also, once strong inversion is reached, W sensitivity tends to vanish. Such an effect is a cancellation between the $1/\sqrt{W}$-law decrease of $\sigma(\Delta V_{TO})$ and the $\sqrt{W}$-law increase of the gate sensitivity to V_{TO} according to Table 2.1.

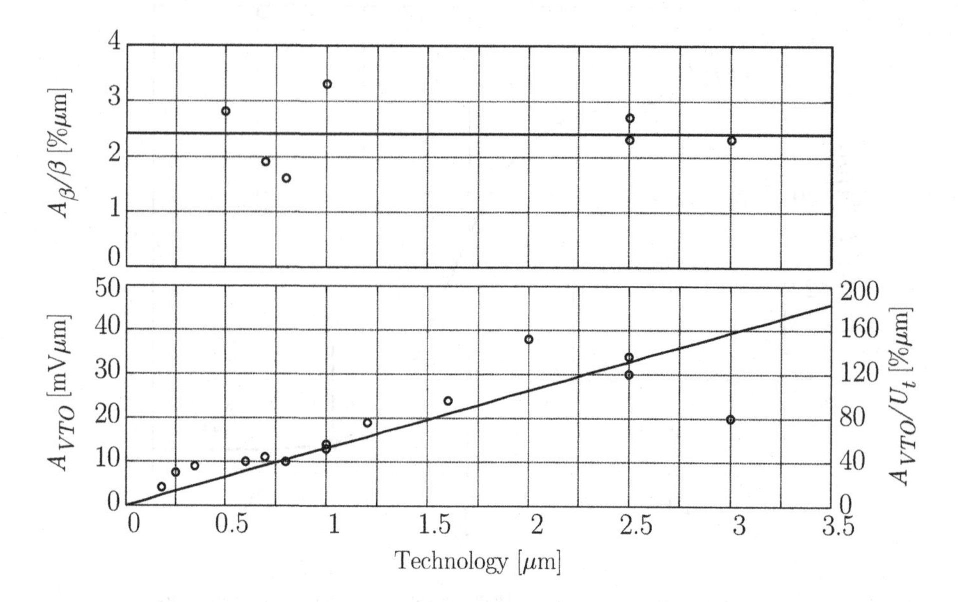

Figure 2.8. NMOSFET β and V_{TO} deviations versus technology generation at room temperature.

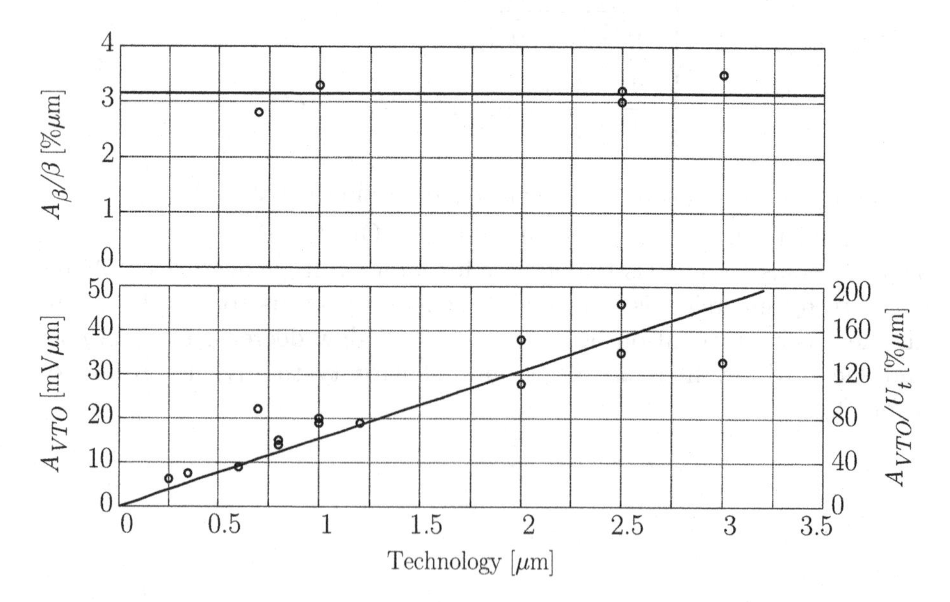

Figure 2.9. PMOSFET β and V_{TO} deviations versus technology generation at room temperature.

6. Parameter Extraction Procedure

In the previous sections, an analytical device model has been set up for the synthesis of the CMOS analog companding circuit techniques developed in this work. However, this by-hand design tool would not be of much benefit without a method to determine its parameter values (i.e. n, β, V_{TO} and λ) for any given CMOS process. The aim of this section is to present a stand-alone procedure to obtain this set of basic parameters either from numerical models or experimental data. This strategy does not intend at all to be an exhaustive device characterization method , but to provide the designer with some independence from the technology models supplied by the foundry.

The procedure starts by extracting the CLM parameter λ from a linear fitting of I_D versus V_{DB} curve in weak and strong inversion saturation regions. The n parameter is computed from the exponential slope of I_D versus V_{GB} in weak inversion saturation, while the unitary current factor (β_u) is extrapolated by applying a quadratic fitting of the same curve in deep strong inversion. Finally, V_{TO} can be extracted from an operation point located in the moderate inversion region of the same I/V characteristics. The complete algorithm is detailed in Table 2.6 and has been numerically implemented through Matlab© [10]. The boundaries of the different regions of operation are automatically selected by studying the curve slopes through the small signal parameters g_{mx}. As can be seen, three curves are needed for the complete extraction, although I_D vs V_{SB} is also checked only for verification. The following examples are based on the same 1.2μm CMOS double-metal double-poly-Si process of all design examples and the industrial application of Chapter 8. The fitting results when applying Table 2.6 (with $\varepsilon = 10\%$ and $V_{DD} = 1.5$V) to experimental NMOSFET curves are shown in Figure 2.10. The equivalent extraction for the typical BSIM3 model is also plotted in Figure 2.11.

Since the analytical equations do not take into account important second-order non-idealities, it is advised to repeat the above procedure for different device geometries in order to study small geometries effects. A matrix device example for the same 1.2μm CMOS process is depicted in Figure 2.12. The resulting exhaustive extraction can be seen in Figures 2.13 and 2.14. The following aspects should be noted:

- Reduction in β for short channels due to velocity saturation at high drain-to-source voltage bias.

- Equivalent increase of V_{TO} for narrow devices caused by lateral bird-peak geometries.

Table 2.6. Procedure for extracting n, β, V_{TO} and λ MOSFET parameters.

Require numerical resolution ε [%], supply voltage V_{DD}.

Require unitary I_D vs V_{DB} in strong inversion:
$I_D(V_{GB0} > V_{TO}, V_{SB0} = 0, V_{DB}) \quad \forall \quad V_{DB} \quad \epsilon \quad [0, V_{DD}]$

Select forward saturation boundaries:

$$V_{DB1} < V_{DB} < V_{DB2} \quad / \quad \frac{g_{md}}{I_D} \geq max\left(\frac{g_{md}}{I_D}\right) \times (1 - \frac{\varepsilon}{100})$$

Compute λ_{si} from linear regression following (2.8):

$$\lambda_{si} = \frac{I_{D2} - I_{D1}}{V_{DB2}I_{D1} - V_{DB1}I_{D2}}$$

Require unitary I_D vs V_{DB} in weak inversion:
$I_D(V_{GB0} < V_{TO}, V_{SB0} = 0, V_{DB}) \quad \forall \quad V_{DB} \quad \epsilon \quad [0, V_{DD}]$

Repeat the two previous steps to obtain λ_{wi}.

Require unitary I_D vs V_{GB} in forward saturation:
$I_D(V_{GB}, V_{SB0} = 0, V_{DB0} = V_{DD}) \quad \forall \quad V_{GB} \quad \epsilon \quad [0, V_{DD}]$

Select weak inversion boundaries:

$$V_{GB1} < V_{GB} < V_{GB2} \quad / \quad \frac{g_{mg}}{I_D} \geq max\left(\frac{g_{mg}}{I_D}\right) \times (1 - \frac{\varepsilon}{100})$$

Compute n from exponential regression following Table 2.1:

$$n = \frac{V_{GB2} - V_{GB1}}{U_t \ln\left(\frac{I_{D2}}{I_{D1}}\right)}$$

Select strong inversion boundaries:

$$V_{GB1} < V_{GB} < V_{GB2} \quad / \quad \frac{g_{mg}}{\sqrt{I_D}} \leq min\left(\frac{g_{mg}}{\sqrt{I_D}}\right) \times (1 + \frac{\varepsilon}{100})$$

Compute unitary β from quadratic regression in Table 2.1:

$$\beta_u = \frac{2n}{1 + \lambda_{si}V_{DB0}}\left(\frac{\sqrt{I_{D1}} - \sqrt{I_{D2}}}{V_{GB1} - V_{GB2}}\right)^2$$

Compute V_{TO} in moderate inversion following Table 2.1:

$$V_{TO} = V_{GB}(I_S) - 2nU_t \ln\left(e^{\frac{1}{\sqrt{1+\frac{\lambda_{wi}+\lambda_{si}}{2}V_{DB0}}}} - 1\right)$$

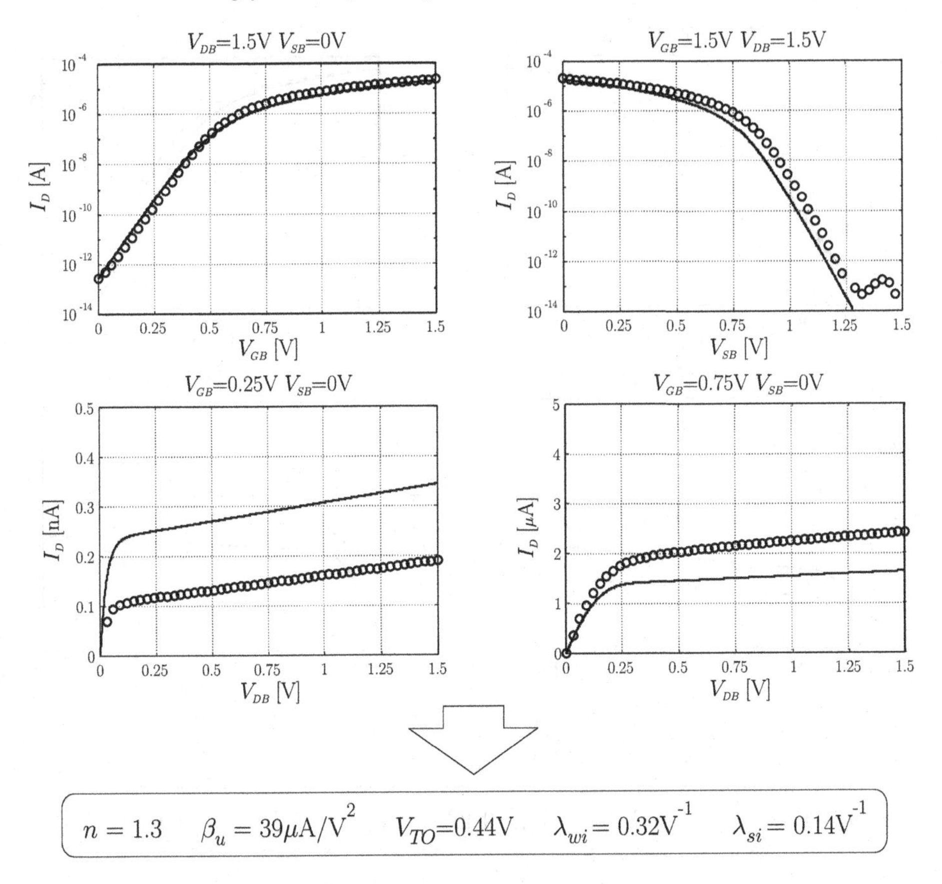

Figure 2.10. Analytical model (solid) extracted from experimental unitary curves (dotted), and fitting results for a (16μm/1.2μm) NMOS device.

- Lower V_{TO} for short MOS transistors originated by the drain and source diffusion charge sharing.
- Inverse proportionality between CLM parameter λ and the channel length itself.
- Good stability of the subthreshold slope n versus both the process spread and the device scaling. In fact, this behavior agrees with the information supplied by the foundry [24] which reports less than a 5% tolerance for 40 lots.

The procedure presented in this section is far from being an exhaustive characterization method [25, 26, 27, 28, 29]. However, its accuracy has been proven good enough to detect practical errors in the selection of BSIM3 parameters, when applied to simulation models, and variations

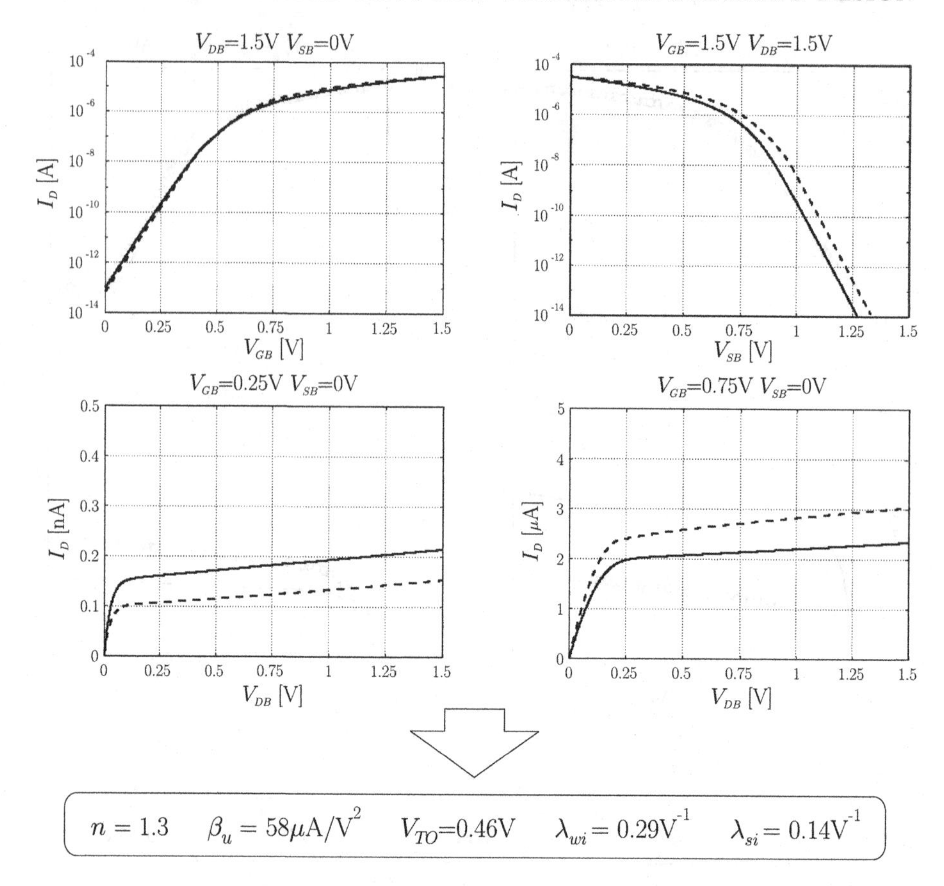

Figure 2.11. Analytical model (solid) extracted from typical BSIM3 unitary curves (dashed), and fitting results for a (16μm/1.2μm) NMOS device.

in electrical parameters of CMOS runs, when used with experimental data. Both situations have been confirmed by the foundry itself [24].

Finally, the extracted MOSFET model must be completed by an estimation of the V_{TO} mismatching parameter. Only a single device geometry is needed to compute the A_{VTO} parameter from (2.24); nevertheless, at least 2 different channel dimensions are recommended to verify the $\frac{1}{\sqrt{WL}}$ law. In our case, statistics for the large and small device geometries of Figure 2.12 were available from the foundry and in-lab measurements, respectively. The correlation between the experimental data and the mismatching model is depicted in Figure 2.15. The resulting $A_{VTO}/U_t \sim 60\%\mu$m agrees with other studies already reported in Section 5. Practical information about numerical simulation of technological mismatching is given in Appendix A.

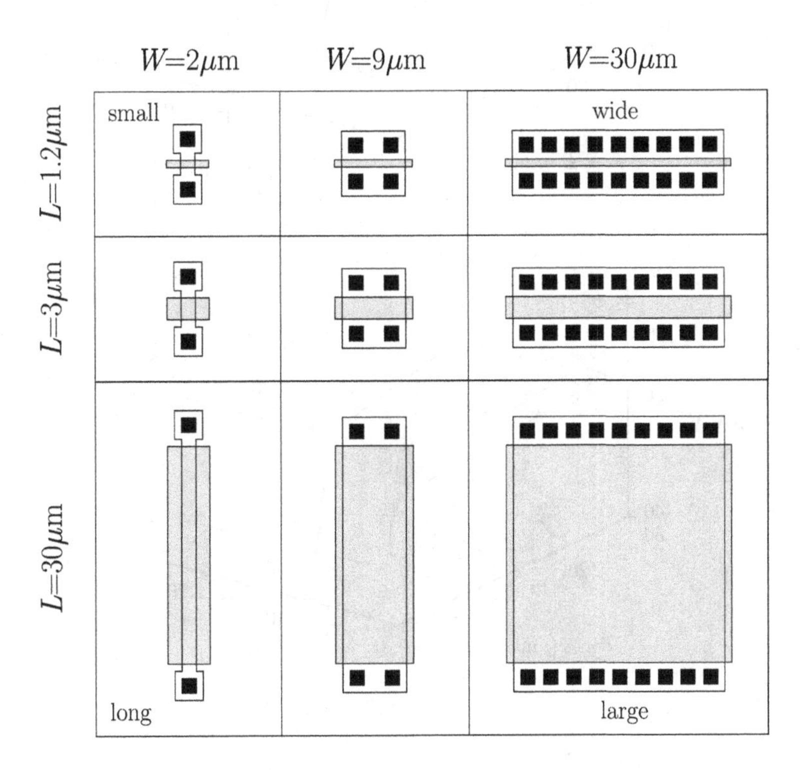

Figure 2.12. Matrix of MOS geometries used in the extraction procedure for a 1.2μm CMOS process example.

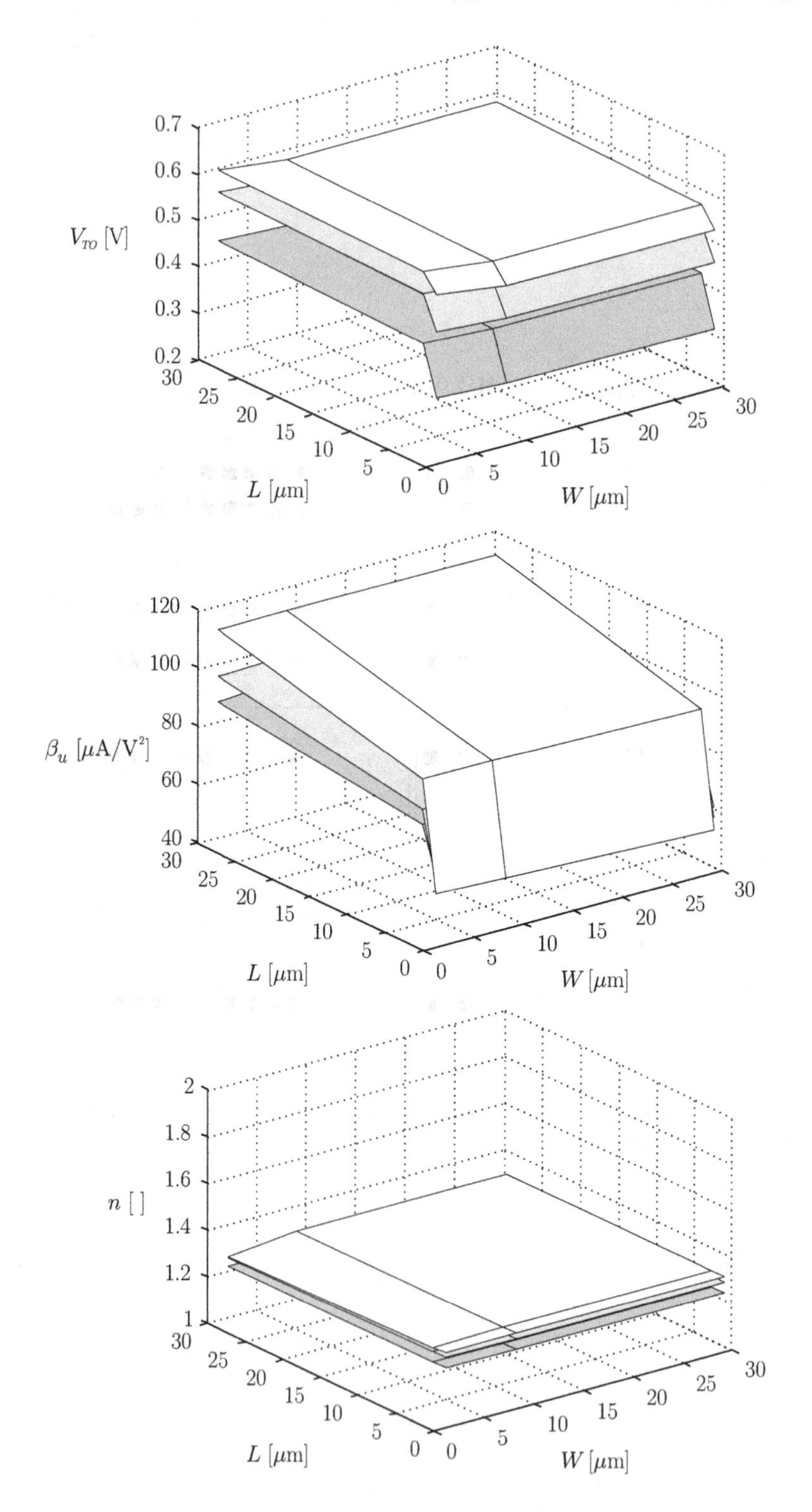

Figure 2.13. Resulting n, β_u and V_{TO} after applying the procedure of Table 2.6 to the typical and corner BSIM3 models of the NMOS devices listed in Figure 2.12.

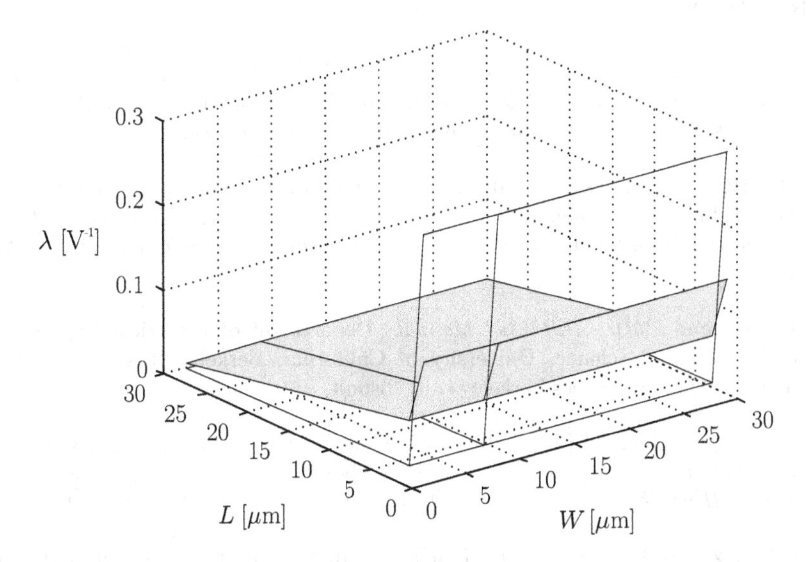

Figure 2.14. Resulting λ in weak (transparent) and strong (solid) inversion after applying Table 2.6 to the typical BSIM3 models of the NMOS devices listed in Figure 2.12.

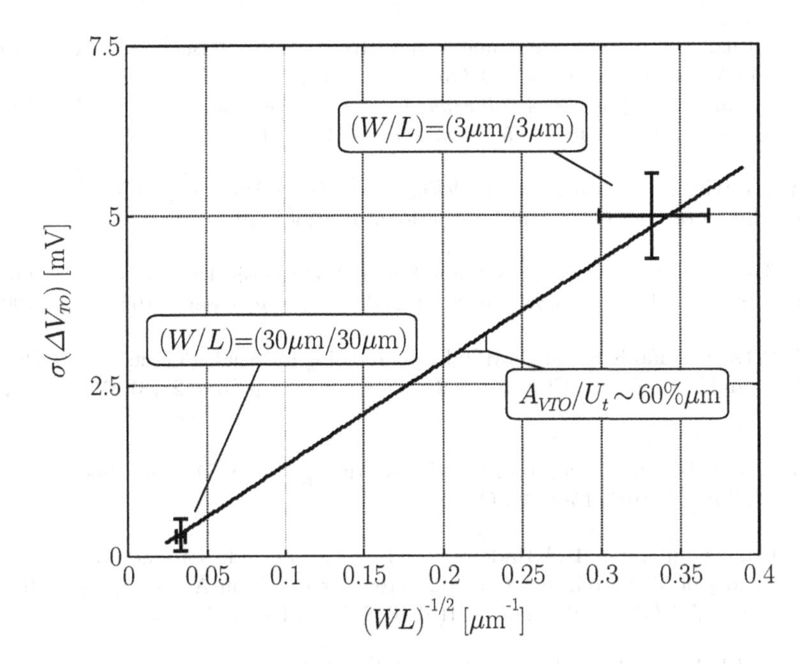

Figure 2.15. Threshold voltage deviations versus MOSFET channel area.

References

[1] Gordon W.Roberts and Adel S.Sedra. *SPICE.* Oxford University Press, New York, 1996. http://www.macs.ece.mcgill.ca/~roberts/spice.

[2] D.Foty. Perspectives on Analytical Modeling of Small Geometry MOSFETs in SPICE for Low Voltage/Low Power CMOS Circuit Design. *Journal of Analog Integrated Circuits and Signal Processing, Kluwer Academic Publishers*, 21:229–252, 1999.

[3] P.K.Ko and C.Hu. *BSIM3v3 Manual.* Department of Electrical Engineering and Computer Science, University of California, Berkeley, CA 94720, 1995. http://www-device.eecs.berkeley.edu/~bsim3.

[4] Y.Cheng, M.Jeng, Z.Liu, J.Huang, M.Chan, K.Chen, P.K.Ko, and C.Hu. A Physical and Scalable I-V Model in BSIM3v3 for Analog/Digital Circuit Simulation. *IEEE Transactions on Electron Devices*, 44(2):277–287, Feb 1997.

[5] C.C.Enz. MOS Translinear Modeling Dedicated to Low-Current and Low-Voltage Analog Circuit Design and Simulation. Technical report, École Politecnique Fédérale de Lausanne, 1995. http://dewww.epfl.ch/leg.

[6] C.C.Enz. *High Precision CMOS Micropower Amplifiers.* PhD thesis, École Politecnique Fédérale de Lausanne, 1989. http://dewww.epfl.ch/leg.

[7] C.C.Enz, F.Krummenacher, and E.A.Vittoz. An Analytical MOS Transistor Model Valid in All Regions of Operation and Dedicated to Low-Voltage and Low-Current Applications. *Journal of Analog Integrated Circuits and Signal Processing, Kluwer Academic Publishers*, 8(1):83–114, 1995.

[8] Y.P.Tsividis. *Operation and Modeling of the MOS Transistor.* Electrical Engineering Series. McGraw-Hill International Editions, 1987.

[9] Y.Sailaja and C.D.Parikh. A Large Signal Non-Quasi-Static Model for Short-Channel MOSFET's. *Physics of Semiconductor Devices*, pages 1048–1051, 1998.

[10] Y.P.Tsividis and K.Suyama. MOSFET Modeling for Analog Circuit CAD: Problems and Prospects. *IEEE Journal of Solid State Circuits*, 29(3):210–216, Mar 1994.

[11] W.Marshall. Fundamentals of Low-Noise Analog Circuit Design. *Proceedings of the IEEE*, 82(10):1515–1538, Oct 1994.

[12] K.R.Lakshmikumar, R.A.Hadaway, and M.A.Copeland. Characterization and Modeling of Mismatch in MOS Transistors for Precision Analog Design. *IEEE Journal of Solid State Circuits*, 21(6):1057–1066, Dec 1986.

[13] M.J.M.Pelgrom, A.C.J.Duinmaijer, and A.P.G.Welbers. Matching Properties of MOS transistors. *IEEE Journal of Solid State Circuits*, 24(5):1433–1440, Oct 1989.

[14] F.Forti and M.E.Wright. Measurement of MOS Current Mismatch in the Weak Inversion Region. *IEEE Journal of Solid State Circuits*, 29(2):138–142, Feb 1994.

[15] Bernabé Linares Barranco. Transistor Mismatch Characterization for CMOS Process CNM 2.5 microns. Registro Provincial de Sevilla 3390, Centro Nacional de Microelectrónica (CNM), Jul 1995.

[16] S.Wong, K.Pan, and D.Ma. A CMOS Mismatch Model and Scaling Effects. *IEEE Electron Device Letters*, 18(6):261–263, Jun 1997.

[17] J.Bastos, M.Steyaert, A.Pergoot, and W.Sansen. Mismatch Characterization of Submicron MOS Transistors. *Journal of Analog Integrated Circuits and Signal Processing, Kluwer Academic Publishers*, 12:95–106, 1997.

[18] U.Grünebaum, J.Oehm, and K.Schumacher. Mismatch Modelling for Large Area MOS Devices. In *Proceedings of the European Solid-State Circuits Conference*, pages 268–271. IEEE, 1997. http://www-be.e-technik.uni-dortmund.de/~grueneba/.

[19] S.J.Lovett, M.Welten, A.Mathewson, and B.Mason. Optimizing MOS Transistor Mismatch. *IEEE Journal of Solid State Circuits*, 33(1):147–150, Jan 1998.

[20] M.J.M.Pelgrom, H.P.Tuihout, and M.Vertregt. Transistor Matching in Analog CMOS Applications. In *Proceedings of the International Electron Devices Meeting*, pages 915–918. IEEE, Dec 1998.

[21] T.Serrano-Gotarredona and B.Linares-Barranco. Systematic Width-and-Length Dependent CMOS Transistor Mismatch Characterization and Simulation. *Journal of Analog Integrated Circuits and Signal Processing, Kluwer Academic Publishers*, 21:271–296, 1999.

[22] J.Oehm, U.Grünebaum, and K.Schumacher. Mismatch Effects Explained by the Spectral Model. In *International Conference on Electronics, Circuits and Systems*, Sep 1999. http://www-be.e-technik.uni-dortmund.de/~grueneba/.

[23] The MathWorks, Inc., 24 Prime Park Way, Natick, MA 01760-1500, USA. *Using MATLAB Version 5.2*, Jan 1998. http://www.mathworks.com.

[24] Austria Mikro Systeme International. Private Communications. http://asic.vertical-global.com, 1998.

[25] M.Bucher, C.Lallement, C.Enz, and F.Krummenacher. Accurate MOS Modelling for Analog Circuit Simulation Using the EKV Model. In *Proceedings of the International Symposium on Circuits and Systems*. IEEE, 1996.

[26] G.A.S.Machado, C.C.Enz, and M.Bucher. Estimating Key Parameters in the EKV MOST Model for Analogue Design and Simulation. In *Proceedings of the International Symposium on Circuits and Systems*, pages 1588–1591. IEEE, 1995.

[27] C.Lallement, M.Bucher, C.Enz, and F.Krummenacher. The EKV MOST Model and the Associated Parameter Extraction. *HP IC-CAP Users Meeting*, 1995.

[28] M.D.Godfrey. CMOS Device Modeling for Subthreshold Circuits. *IEEE Transactions on Circuits and Systems-II*, 39(8):532–539, Aug 1992.

[29] J.Gómez-Cipriano, L.Bitencourt, and S.Bampi. Comparison of the BSIM3 and EKV Model Parameter Extraction Methodologies with a Direct Search Optimization Method. *Proceedings of the III Workshop Iberchip, CINVESTAV-PIN, México D.F.*, Feb 1997.

Chapter 3

AMPLIFICATION AND AGC

Abstract This chapter includes all the new circuit techniques developed for gain stages. After reviewing its Log companding principle, the CMOS core topologies are proposed and compared. The optimum strategy is chosen in terms of low-voltage and CMOS technology compatibility. A complete set of very low-voltage basic building blocks is then presented for either fixed, programmable or syllabic full AGC systems. Finally, the validity of the novel CMOS circuit techniques are demonstrated through some design examples.

1. Log Companding Principle

The main purpose of any amplifying stage is to obtain a linearly scaled copy of the input signal at the output port. Using the generalized nomenclature introduced in Section 3, this proportionality can be specified as:

$$y_{out} \doteq G y_{in} \tag{3.1}$$

where G stands for the gain factor. The above transfer function must then be translated to the compressed x-domain in order to identify the necessary processing after the compressor and before the expander. This step is accomplished by applying the Log companding function F of (1.5) to the previous expression:

$$x_{out} = x_{gain} + x_{in} \tag{3.2}$$

where:

$$G \doteq e^{x_{gain}} \qquad G[\text{dB}] = 20\log(e)x_{gain} \tag{3.3}$$

with x_{gain} being the gain control signal necessary to keep the system externally linear. It is easy to see that external amplification is

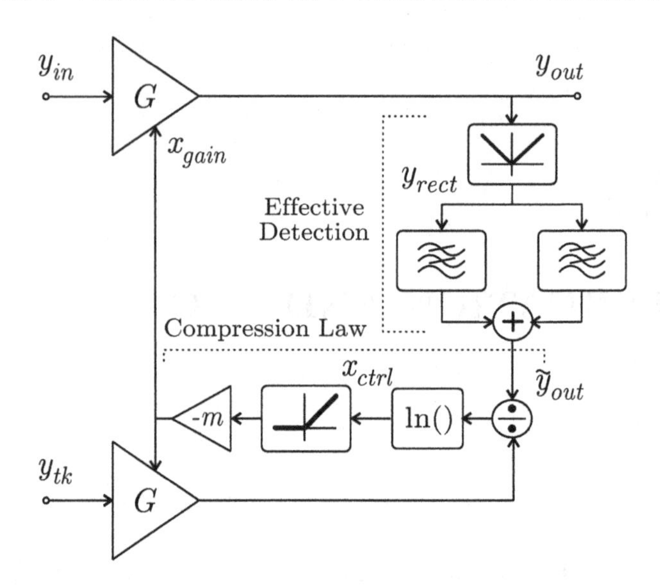

Figure 3.1. General AGC model using Log amplifiers.

equivalent to internally adding an independent term, as indicated in [1, 2]. Direct control of x_{gain} allows electronic tuning for either fixed, programmable or time-variant amplification factors. Furthermore, the linear relation in a log scale such as (3.3) is of special interest when processing signals with large dynamic range (e.g. audio applications). In particular, the use of such logarithmic amplifiers simplifies the synthesis at the system level of compression laws in AGC algorithms [3, 4]. In this sense, a general attenuation system is modelled in Figure 3.1.

The AGC system usually involves two identical Log amplifiers devoted to signal processing (y_{out}) and control of the threshold knee point (y_{tk}), respectively. The feedback path begins with an envelope detector which sets the transient response of the whole loop through the attack and release times [5]. Typically, this block consists on a full-wave rectification (y_{rect}) plus some filtering to compute the effective envelope ($\tilde{y}_{out}$) according to the particular signal processing application. For example, in the case of syllabic AGC for hearing aids, two low-pass filters with time constants around 10ms and 250ms are usually combined to ensure fast protection against overshoots while maintaining speech intelligibility, respectively. The result of comparing $\tilde{y}_{out}$ and the processed y_{tk} is then translated to the Log domain, where first-quadrant-only propagation automatically ensures linear operation for $\tilde{y}_{in} \leq y_{tk}$ (i.e. the threshold knee point). Once in close loop operation, the scaling factor m defines the following non-linear compression curve:

$$\tilde{y}_{out} = \begin{cases} \tilde{y}_{in} & \tilde{y}_{in} \leq y_{tk} \\ y_{tk}^{m} \tilde{y}_{in}^{1-m} & \tilde{y}_{in} > y_{tk} \quad \text{and} \quad 0 < m < 1 \end{cases} \tag{3.4}$$

Hence, the resulting output dynamic range reduction can be expressed using the compression Ratio (CR):

$$CR \doteq \frac{\partial \tilde{y}_{in}[\text{dB}]}{\partial \tilde{y}_{out}[\text{dB}]} = \frac{1}{1-m} \qquad 0 < m < 1 \tag{3.5}$$

where upper and lower boundaries from (3.4) are the limiting (i.e. $CR = \infty : 1$, $m = 1$, $\tilde{y}_{out} \equiv y_{tk}$) and linear (i.e. $CR = 1 : 1$, $m = 0$, $\tilde{y}_{out} \equiv \tilde{y}_{in}$) cases, respectively. Note that AGC compressing must not be confused with the companding approach itself, since the latter operates instantaneously, restores the original dynamic range at the output, and is directly synthesized at the transistor level.

2. CMOS Generalization

As was mentioned in the introduction of this work, the circuit implementation of the general Log companding function (1.5) at the device level will be performed here by the use of the MOSFET I/V curves in the weak inversion region. Furthermore, since the basic amplifier block can be formally understood as a current-controlled current source with voltage-controlled exponential gain, saturation operation is chosen here for the main MOS transistors in order to obtain a higher output impedance. Taking the asymptotic expression of I_D in Table 2.1 for the forward case and neglecting channel length modulation (CLM), three terminals of the MOSFET exhibit the desired exponential law F: gate, source and bulk. The resulting internal signal compression through each terminal will be named gate- (GD), source- (SD) and bulk-driven (BD), respectively:

$$I = F(V) = \begin{cases} I_S e^{-\frac{V_{TO}+nV_{bias}}{nU_t}} e^{\frac{V}{nU_t}} & \text{GD} \\ I_S e^{\frac{V_{bias}-V_{TO}}{nU_t}} e^{-\frac{V}{U_t}} & \text{SD} \\ I_S e^{\frac{V_{bias}-V_{TO}}{nU_t}} e^{\left(1-\frac{1}{n}\right)\frac{V}{U_t}} & \text{BD} \end{cases} \tag{3.6}$$

where V, I and V_{bias} stand for the compressed voltage signal at the selected terminal, the linear current signal at the drain, and the biasing reference for the other terminals, respectively. Based on the required processing in the compressed V-domain from equation (3.2), three different G tuning techniques are also proposed: gate- (GC), source- (SC) and bulk-controlled (BC), which correspond respectively to the V_{gainGC}, V_{gainSC} and V_{gainBC} sources in the summary of Figure 3.2. Both de-

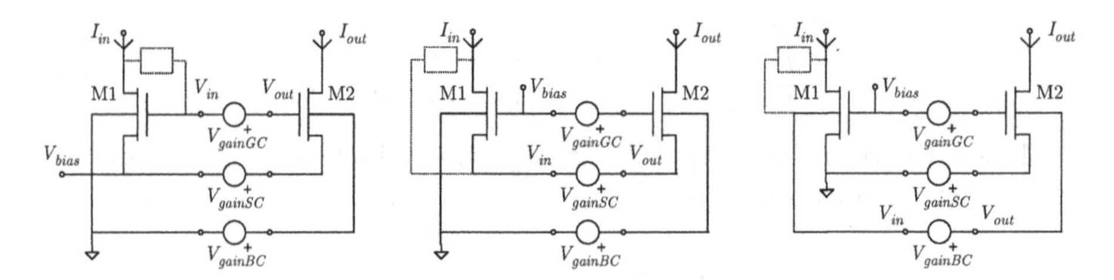

Figure 3.2. Summary of gain controlling topologies for GD (left) SD (center) and BD (right) realizations (auxiliary circuitry in dashed).

vices M1 and M2 are supposed to operate in weak inversion saturation in order to implement the compression and the expansion functions.

In fact, all three types of gain control can be mixed together and are compatible with all three types of driven signals in (3.6). Supposing technological, geometrical and thermal matching between M1 and M2 devices, the unified expression of the resulting gain G in the I-domain is given by:

$$G = \frac{I_{out}}{I_{in}} = e^{\frac{V_{gainGC}}{nU_t}} e^{\frac{-V_{gainSC}}{U_t}} e^{\frac{n-1}{n}\frac{V_{gainBC}}{U_t}} \tag{3.7}$$

$$G[\mathrm{dB}] = 20\log(e)\left[\left(\frac{1}{n}\right)\frac{V_{gainGC}}{U_t} - \frac{V_{gainSC}}{U_t} + \left(\frac{n-1}{n}\right)\frac{V_{gainBC}}{U_t}\right] \tag{3.8}$$

The first conclusion from the above equations is the verification of the linear relation in decibels between G and all gain control voltages. Another important feature of all strategies is the possibility of synthesizing both amplification and attenuation factors too, depending on the sign of the differential voltage V_{gain}. In the trivial case of $V_{gain} = 0$, all topologies are reduced to the classical current mirror ($G \equiv 0$dB). Also, all control strategies require a proportional-to-absolute-temperature (PTAT) voltage reference to cancel their first-order thermal sensitivity due to U_t. The synthesis of such references is addressed in Chapter 5.

However, there are some important differences between gate-, source- and bulk-controlled topologies which force the choice of the optimum solution:

GC does not need low-ohmic V_{gainGC} sources, although some technological dependency still remains through parameter n.

BC approach exhibits the largest control voltages for the normal range of $1 < n < 2$, causing an extreme reduction in the available room for

Table 3.1. Gain tuning factors versus topology for $n = 1.3$.

$\lvert G\rvert$ [dB]	$\lvert V_{gainGC}\rvert$ $[U_t]$	$\lvert V_{gainSC}\rvert$ $[U_t]$	$\lvert V_{gainBC}\rvert$ $[U_t]$
1	0.15	0.11	0.50
20	3.0	2.3	10
40	6.0	4.6	20
60	9.0	6.9	30

low-voltage operation and incompatibility with anti-latch-up rules of standard CMOS processes. For example, a $G = \pm 60$dB would need a $V_{gainBC} \sim \mp 750$mV for $n = 1.3$ at room temperature.

SC topology shows the best gain sensitivity versus control voltage, as seen in Table 3.1, and independence from technology as well. Unfortunately, it requires low-impedance sources for the control of V_{gainSC}.

3. Basic Building Blocks

This section proposes very low-voltage CMOS implementations of all the required auxiliary circuitry around Figure 3.2 in order to build up from programmable amplifying stages to complete AGC systems in the Log domain.

3.1 General-Purpose Controllable Amplifier Cell

Taking into account all the considerations explained in the previous section, the GD-SC combination is chosen here as the most suitable topology, so the rest of this chapter will be devoted only to this approach. The differential control voltage V_{gainSC} in Figure 3.2 splits into two sources referred to as ground ($V_{gaini,o}$) as shown in Figure 3.3. Unless specified, local substrates of NMOS and PMOS devices in all circuits are assumed to be connected to minimum (V_{SS}) and maximum (V_{DD}) system potentials, respectively.

This modification simplifies the synthesis of such low-impedance sources, as discussed later on in this section, and also supplies two independent signals in the Log domain for gain control, as in the AGC example of Section 4. From (3.8), the resulting gain expression is:

$$G[\text{dB}] = 20\log(e)\frac{V_{gaini} - V_{gaino}}{U_t} \tag{3.9}$$

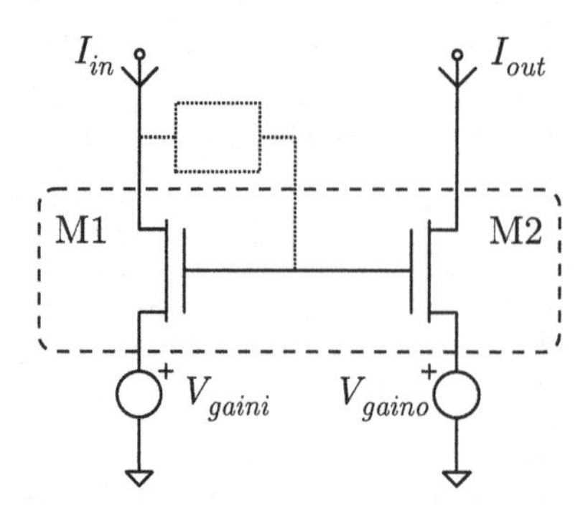

Figure 3.3. Schematic of the GD-SC NMOS cell (auxiliary circuitry in dashed).

with sensitivity around 0.43dB/mV at room temperature. In practice, these gain control signals are usually limited to $0 < V_{gain} < 7U_t$, which corresponds to a total programmable amplification range as large as $\pm$60dB.

As with any current processor, some extra circuitry must be added to Figure 3.3 in order to provide a low enough input impedance, ideally null. However, zero impedance cannot be achieved in practice. Furthermore, dynamics has to be introduced at this point because even low enough values of input impedance are difficult to keep so low along the entire frequency spectrum due to the parasitic capacitance present in any CMOS environment. In particular, the input capacitance (C_{in}) will be an important bandwidth limiter, equivalent to the output load capacitance in voltage processing. The simplest circuit solution for the control of the M1 compressor is shown in Figure 3.4(left), where I_{biasi} stands for the input biasing. However, the active load topology exhibits a strong relation between the input impedance and I_{biasi} itself through $1/g_{mg1}$, which in practice does not achieve low enough values. As a result, CLM effects tend to cause input distortion at the compressor stage. Hence, the optimum strategy aims to minimize the drain voltage slew at M1.

The above circuit problem has already been faced in the design of dynamic current copiers for Switched-current (SI) systems, as in [6]. The solution classically proposed for low-voltage operation [7, 8, 9] makes use of a voltage gain stage inserted between drain and gate of the input transistor M1, symbolized in Figure 3.4(right) by an operational voltage amplifier (OVA). Indeed, the static input impedance is reduced by the DC voltage gain of the OVA block (G_{OVA}), and so distortion, due to CLM. On the other hand, as warned by some authors [10, 11], a drawback arises when trying to stabilize this local feedback loop, split now into two poles. In order to identify the exact cause of such instability, a complete

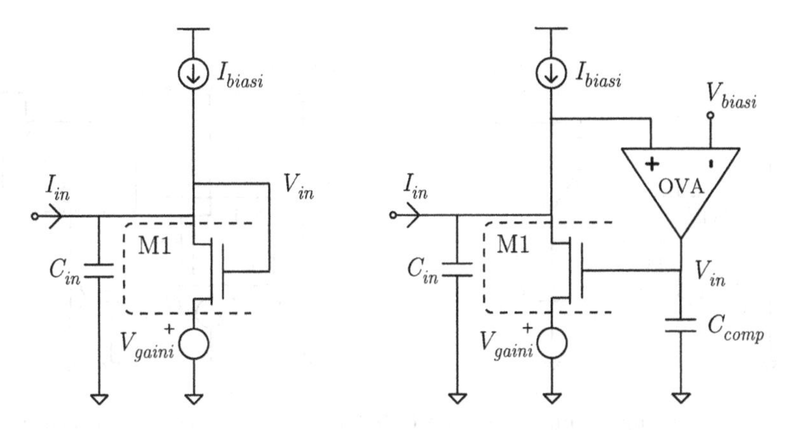

Figure 3.4. Simple active load (left) and classic input impedance control (right).

small-signal analysis is applied to study both the input impedance (Z_{in}) and the transfer function (T):

$$Z_{in}(s) \simeq \frac{1}{Lg_{md1}} \frac{\frac{s}{p_{comp}} + 1}{\frac{s^2}{p_{in}p_{comp}L} + \frac{p_{in}+p_{comp}}{p_{in}p_{comp}} s + 1} \qquad L \gg 1 \tag{3.10}$$

$$T(s) \simeq \frac{1}{g_{md1}} \frac{1}{\frac{s^2}{p_{in}p_{comp}L} + \frac{p_{in}+p_{comp}}{p_{in}p_{comp}} s + 1} \qquad L \gg 1 \tag{3.11}$$

where L, p_{in} and p_{comp} stand for the loop gain, input and compensation poles, respectively:

$$L \doteq \frac{g_{mg1}}{g_{md1}} G_{OVA} \qquad p_{in} \doteq \frac{g_{md1}}{C_{in}} \qquad p_{comp} \doteq \frac{g_{outOVA}}{C_{comp}} \tag{3.12}$$

As predicted, the static input impedance $Z_{in}(DC) = \frac{1}{Lg_{md1}} = \frac{1}{g_{mg1}G_{OVA}}$ has been reduced by the OVA gain factor (G_{OVA}). However, from a dynamic point of view, the new two-pole distribution of both $T(s)$ and $Z_{in}(s)$ returns the following Damping factor (ζ):

$$\zeta \equiv \frac{1}{2}\sqrt{\left(\frac{p_{in}}{p_{comp}} + \frac{p_{comp}}{p_{in}}\right)\frac{1}{L}} \simeq \frac{1}{2}\sqrt{\frac{p_{in}/p_{comp}}{L}} \qquad p_{comp} \ll p_{in} \tag{3.13}$$

Since in practice the OVA block must also be implemented by means of CMOS devices, its output impedance will be in the same order as the input transistor M1. Thus, ζ can then be expressed only in terms of capacitance ratios:

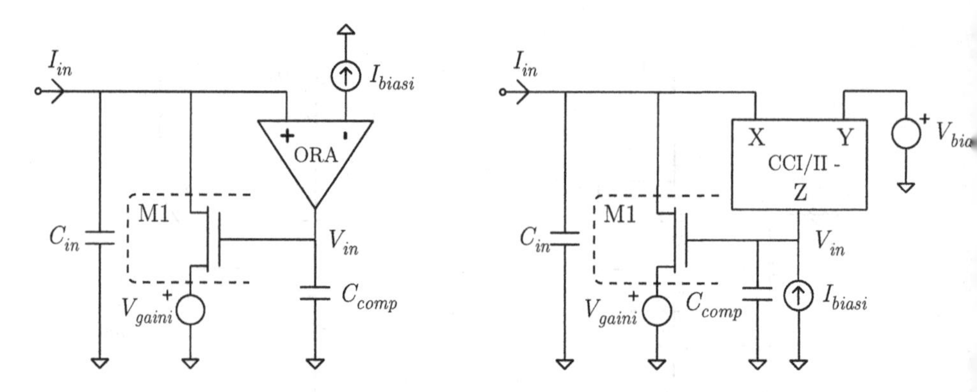

Figure 3.5. New input impedance control (left) and its CCI/II- equivalence (right).

$$\zeta \simeq \frac{1}{2}\sqrt{\frac{C_{comp}}{C_{in}L}} \qquad g_{md1} \sim g_{outOVA} \tag{3.14}$$

Now the stability problem can be easily understood: a large pole splitting is required for practical values of $L \gg 1$ to limit the closed loop frequency overshoot. Note that L not only includes G_{OVA} but also the gain of the equivalent feedback inverter M1-I_{biasi}. The more G_{OVA} is increased to obtain lower input impedance, the more capacitance ratio is needed to ensure stability. For example, taking the general expression of the maximum frequency overshoot:

$$\Delta T_{max} = \frac{1}{2\zeta\sqrt{1-\zeta^2}} \qquad \zeta < \frac{1}{\sqrt{2}} \tag{3.15}$$

flat response ($\zeta > \frac{1}{\sqrt{2}}$) would need a C_{comp}/C_{in} ratio of about two decades for a minimum $L >$ 40dB! In practice, the compensation capacitor C_{comp} for a minimum realistic value of parasitic C_{inp} would not be CMOS integrable and would also drastically reduce the available signal bandwidth at the input compressor.

A novel circuit technique is presented in Figure 3.5 to lower the input impedance while relaxing stability conditions, too. The new strategy makes use of an operational transresistance amplifier (ORA) in the feedback loop. The equivalent description in terms of current Conveyors [12, 13, 14] is depicted in the same figure to clarify the functionality of this block.

The circuit ensures the DC biasing of M1 at I_{biasi} and signal propagation as the ORA tends to balance its input currents by controlling the gate of the input transistor. However, unlike the classical approach, the ideally null input impedance of the ORA moves the input pole to high frequencies ($p_{in} \rightarrow \infty$) while reducing the voltage loop gain ($L \rightarrow 0$).

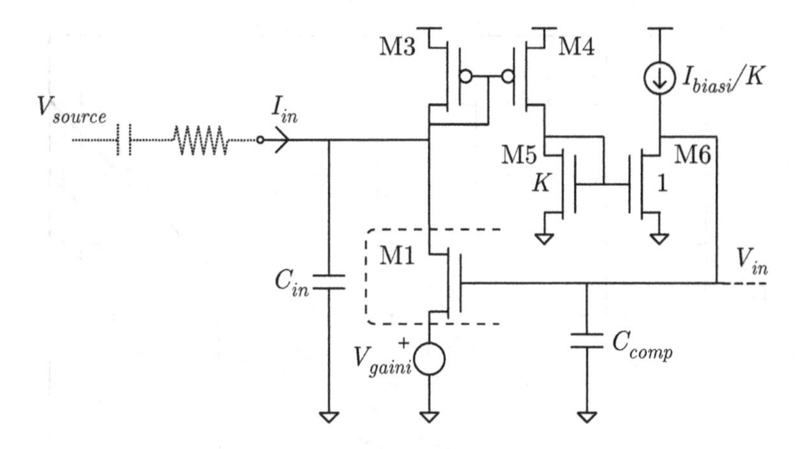

Figure 3.6. Low-voltage CMOS implementation proposal for Figure 3.5.

The compensation pole, however, can be easily kept at the same location by making $g_{outORA} \sim g_{outOVA}$. As a result, the stability condition (3.13) is relaxed to integrable values of C_{comp}. Based on this strategy, a low-voltage CMOS implementation is proposed in Figure 3.6, where the ORA block is built through M3-M6. The key parameters for the stability condition can be rewritten as follows:

$$L \simeq \frac{1}{K}\frac{g_{mg3}}{g_{md6}} \qquad p_{in} = \frac{g_{mg3}}{C_{in}} \qquad p_{comp} = \frac{g_{md6}}{C_{comp}} \tag{3.16}$$

Apart from its good low-voltage compatibility, a particular advantage of Figure 3.6 is the fact that the pole splitting exhibits a direct proportionality to the voltage loop gain:

$$\frac{p_{in}}{p_{comp}} \equiv \frac{KC_{comp}}{C_{in}}L \tag{3.17}$$

In consequence, the stability condition becomes affordable and easier to design by hand calculation:

$$\zeta = \frac{1}{2}\sqrt{\frac{KC_{comp}}{C_{in}}} \tag{3.18}$$

Recalling the same example mentioned before, flat response ($\zeta > \frac{1}{\sqrt{2}}$) requires in this case only a suitable dimensioning of the compensation capacitor $C_{comp} > 2C_{in}/K$. The graphical comparison of Figure 3.7 demonstrates the validity of this circuit strategy:

Finally, the additional design constraint to choose the absolute values of K and C_{comp} can be derived from bandwidth requirements:

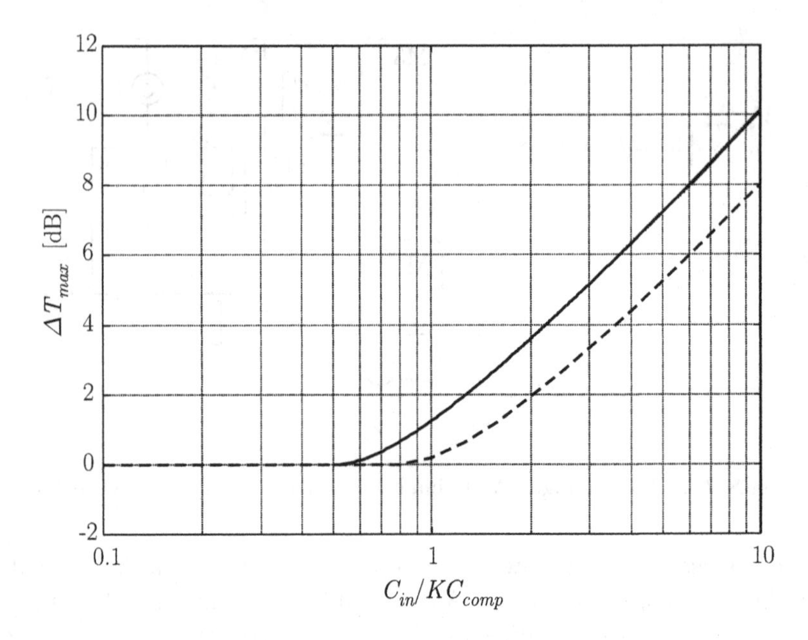

Figure 3.7. Analytical (solid) and BSIM3 simulated (dashed) maximum frequency overshoot versus normalized compensation.

$$f_{-3\mathrm{dB}} = \frac{p_{comp}}{2\pi} L \sqrt{\frac{KC_{comp}}{C_{in}}} \tag{3.19}$$

The following flow diagram depicts the complete design process:

$$\begin{array}{ccccc} C_{inmax} & \xrightarrow{(3.18)} & (KC_{comp})_{min} & \xrightarrow{(3.19)} & K \text{ and } C_{comp} \\ & & \uparrow{\scriptstyle (3.18)} & & \\ \Delta T_{max} & \xrightarrow[(3.15)]{} & \zeta_{min} & & \end{array} \tag{3.20}$$

Due to the resulting impedance values (typically below 1KΩ for $I_{biasi} \sim 1\mu$A), optional V/I conversion at the input linear V-domain can be easily obtained by a simple series of a resistor and a decoupling capacitor, as illustrated in Figure 3.6. Furthermore, the proposed control technique minimizes the CLM effect at the compressor M1, allowing even minimum channel length selection for such a device. This possibility means an important Si area saving due to the wide aspect ratios usually required in compressors and expanders. The same technique also applies to the expander transistor M2, provided that a similar control loop is included at the input of the cascaded stage.

Table 3.2. Automatic biasing levels.

	$G < 1$	$G > 1$
$\frac{I_{biasi}}{I_{max}}$	1	$\frac{1+N}{1+NG}$
$\frac{I_{biaso}}{I_{max}}$	$\frac{1}{G}$	$\frac{1+N}{1+NG}G$

The next design step consists of adding the biasing scheme to the basic cell of Figure 3.3. This task includes the generation of both the input bias level (I_{biasi}) to accommodate the incoming linear signal (I_{in}), and the proper value to be subtracted at the expander (I_{biaso}) in order to obtain a DC-free output signal (I_{out}). Also, the auxiliary circuitry dedicated to biasing must adjust such current references dynamically to any changes in the states of $V_{gaini,o}$ caused by either user programming or AGC feedback. In this sense, the key design parameter here is the maximum signal level or full-scale (I_{max}) according to the power and dynamic range issues explained in Chapter 7.

A general low-voltage CMOS implementation of the biasing scheme is proposed in Figure 3.8. The proper DC current levels are computed here through M7-M8 which act as a parallel GD-SC cell controlled by the same gain G and operated in principle as an attenuator (i.e. $I_{biasi} = I_{max}$). Its output is permanently monitored by M9-M10, so in case it exceeds the allowed range (i.e. $I_{biaso} > I_{max}$), the error amplifier M11-M12 automatically corrects the input I_{biasi} value. The required copies of both $I_{biasi,o}$ for the compressor and the expander devices M1 and M2 are obtained through M13-M15 and M16 respectively. Hence, the proposed solution is compatible either with amplification ($G > 1$) or attenuation ($G < 1$) factors. The detailed biasing expressions for each case are given in Table 3.2, where the design variable N sets the resolution of the whole control. For $G > 1$, larger N values return better insensitivity to G as can be seen in Figure 3.9. In these cases, some frequency compensation at M7-M8 may be required when $G \gg 1$.

3.2 Low-Impedance Gain Control Voltage Sources

Once the signal path of the basic GD-SC amplifier cell has been completed, this subsection is addressed to the control of its gain G. Even for fixed gain stages, built-in voltage-controlled voltage sources are usually needed to translate a general high-impedance PTAT reference or control signal (V_{ctrl}) to the desired low-impedance V_{gain} port of the amplifier in

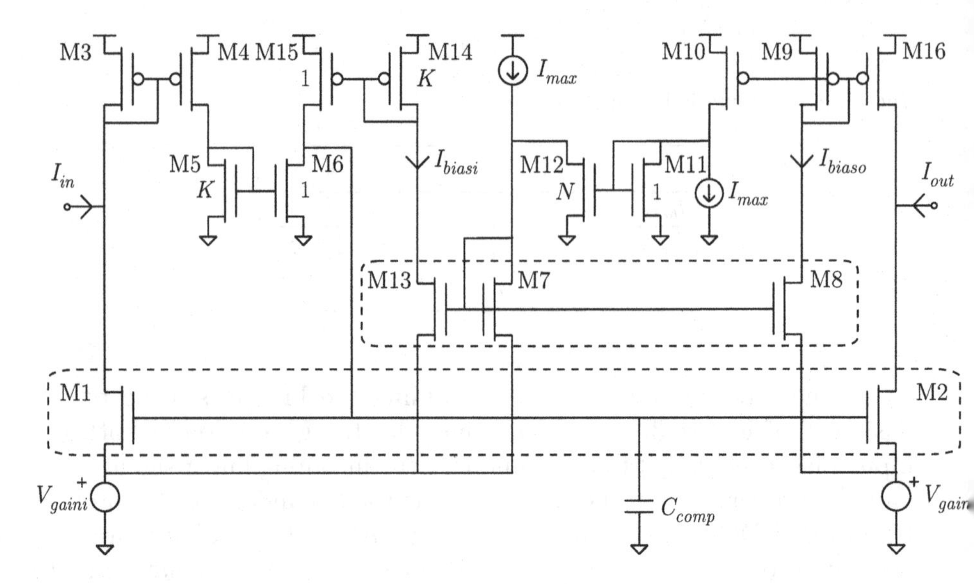

Figure 3.8. Low-voltage implementation of the controllable amplifier cell.

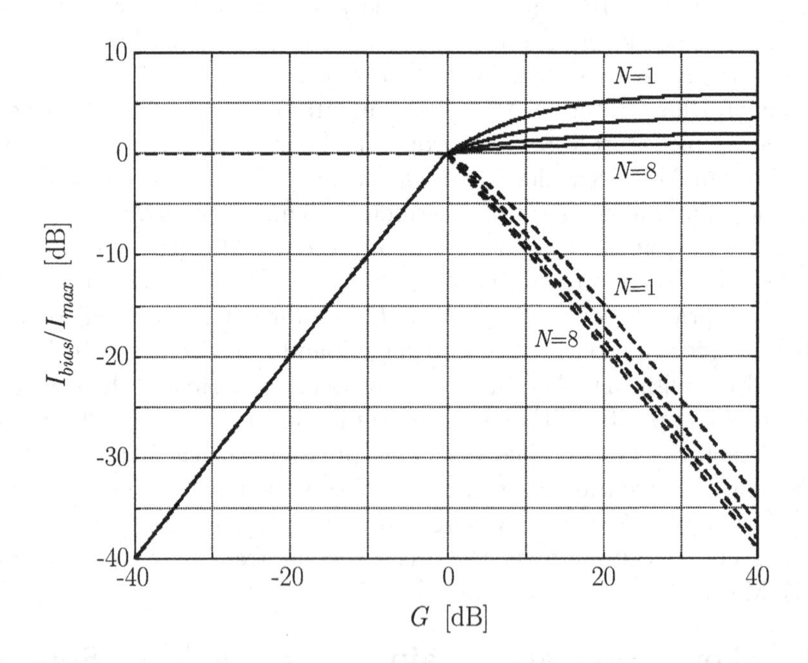

Figure 3.9. Normalized input (dashed) and output (solid) auto-biasing values for feedback factors $N = 1$, 2, 4 and 8.

Figure 3.8. Due to the inner voltage compression of the GD-SC topology, V_{gain} signals are commonly limited to less than $8U_t$ (i.e. 200mV at room temperature), so no rail-to-rail operation is needed for such voltage sources. In fact, the main circuit specifications are:

- Low-amplitude input ($0 < V_{ctrl} < 8U_t$).
- High-impedance input ($I_{ctrl} = 0$).
- Follower output ($V_{gain} = V_{ctrl}$).
- Low-impedance output with sink capability ($0 < I_{gain} < I_{max}$).

Previous implementations for this kind of auxiliary blocks are based on a constant-current-fed resistor [2, 15]. However, apart from requiring the corresponding resistive layer in the CMOS process, this approach suffers from a bad compromise between power and area. On the one hand, the static consumption of the source must be set as high as I_{max}. On the other hand, such a value is dependent on the V_{ctrl} itself and the resistor value. As a result, this circuit solution exhibits an inverse proportionality between power consumption and the resistor value, which usually translates into a high Silicon area. In the case of minimizing the static current consumption of the source, the value of the resistor must be increased to reach V_{ctrl} specifications. On the other hand, if compact solutions are preferred by scaling down the resistor element, the quiescent current needs to be multiplied by the same factor to obtain the desired output voltage.

A new circuit alternative is proposed here based on a variable MOS resistor. The basic strategy is depicted in Figure 3.10, where M1 plays the role of the tunable resistor element operating in strong inversion conduction. The telescopic biasing through M2 ensures proper operation even for almost-zero values of V_{gain}. The OVA block supplies an additional gain (G_{OVA}) in the case of that very low ohmic outputs are required. In fact, and since M2 is operating in saturation, the general expression of the output impedance can be approximated by:

$$r_{out} = \frac{1}{g_{md1} + g_{mg1}\left(1 + \frac{g_{ms2} + G_{OVA} g_{mg2}}{g_{md2}}\right)} \simeq \frac{1}{g_{mg1}\frac{g_{mg2}}{g_{md2}}(n + G_{OVA})} \tag{3.21}$$

Two different low-voltage CMOS implementations based on this telescopic topology are presented in Figure 3.11. The first solution (left) is optimized against technology mismatching by using the reduced group of devices M3-M6 to compute the error signal. Furthermore, this option returns a very low output impedance at the V_{gain} port while preserving a high input impedance at the input V_{ctrl}. However, only moderate

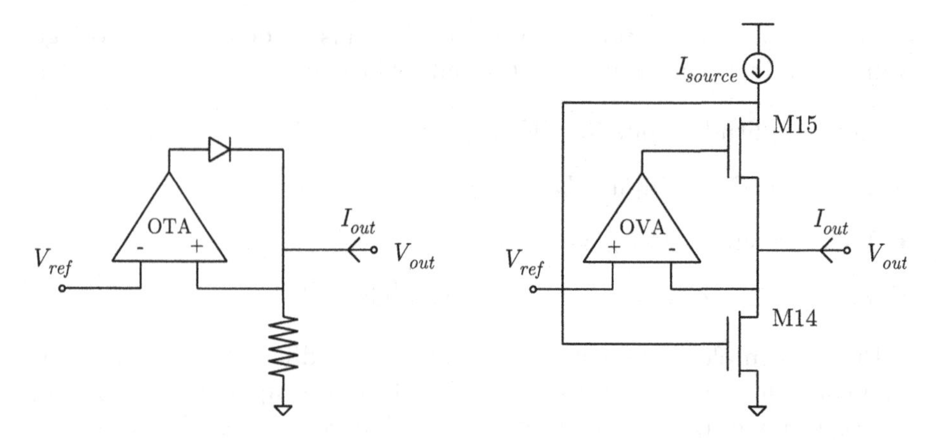

Figure 3.10. General low-ohmic model (left) and low-voltage topology (right) proposed for V_{gain} sources.

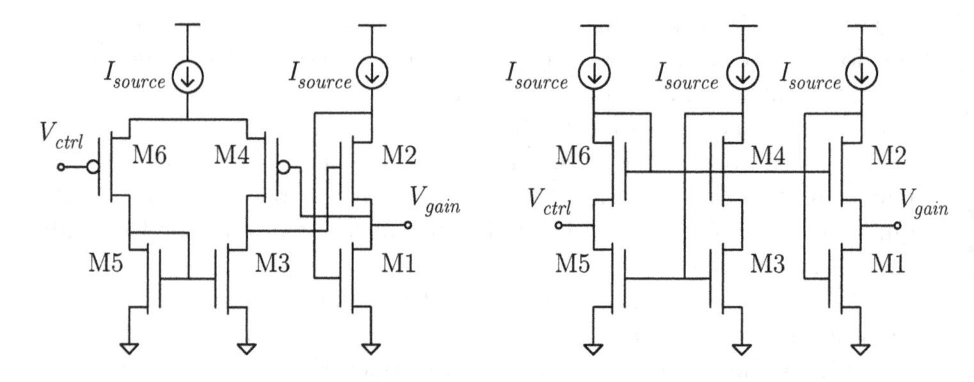

Figure 3.11. Low-voltage CMOS controlled sources proposed for low technology mismatching (left) and low output noise (right).

noise performance is achieved due to the gain factor G_{OVA} synthesized by the same four devices M3-M6. In this sense, a second approach is also proposed in Figure 3.11(right) to alleviate this last problem. Although not very low or high impedance can be obtained at the output and input ports respectively, an important resolution improvement comes from the fact that $G_{OVA} \equiv 0$.

In any case, in both cases I_{source} should be chosen according to the bandwidth and Power Supply Rejection ratio (PSRR) requirements of the particular application. Another interesting design parameter for this basic building block is its minimum output voltage ($V_{gainmin}$). Taking the expression of the drain current in strong inversion conduction for M1 from Table 2.1, the overload situation will occur when:

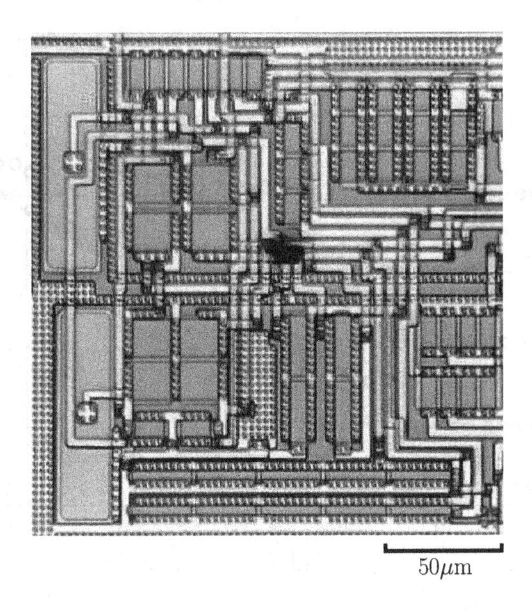

Figure 3.12. Microscope photography of a dual gain controlled source corresponding to the proposal of Figure 3.10 (left).

$$I_{max} + I_{source} = \beta_1 \left[(V_{DD} - V_{sat} - V_{TO}) - \frac{n}{2} V_{gainmin} \right] V_{gainmin} \quad (3.22)$$

where V_{sat} stands for the minimum voltage drop of the output I_{source} to ensure its nominal value and good PSRR. Supposing large sinking ($I_{max} \gg I_{source}$), the final design equation can be expressed as:

$$V_{gainmin} \simeq \frac{I_{max}}{\beta \left(\frac{W}{L}\right)_1 (V_{DD} - V_{sat} - V_{TO})} \quad (3.23)$$

An integrated realization of the proposed low-impedance gain control voltage source of Figure 3.11(left) is depicted in Figure 3.12, while Figure 3.13 shows its typical behaviour.

3.3 Full-Wave Rectifiers

In the case of building AGC systems around the proposed Log amplifier, such as the general model of Figure 3.1, rectifying is usually the first step towards the computation of the approximate root-mean-square (RMS) value of the output signal. Furthermore, full-wave rectifiers are preferable in order to minimize the ripple in the control path of the AGC, otherwise higher order filtering is required to avoid excessive output distortion. In practice, a key design parameter for the rectifier is

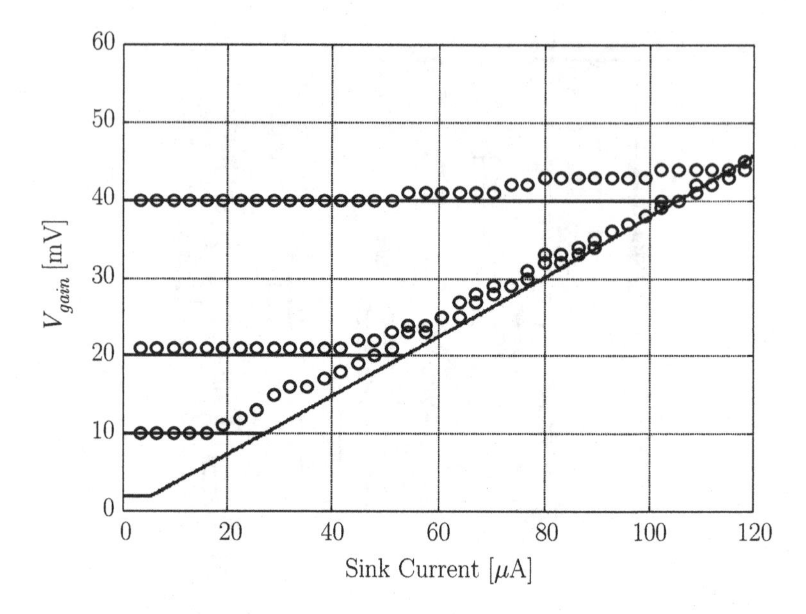

Figure 3.13. Experimental (dotted) and analytical (solid) output control voltage versus sink current for Figure 3.12. Design parameters are $V_{gainmin} \doteq 2$mV, $I_{max} \doteq 4\mu$A and $I_{source} \doteq 1\mu$A. Experimental quantization error is about 1mV.

the maximum output error in absence of an input signal, which gives the effective resolution of the rectifying process. Such an offset is usually generated by a combination of two sources: a residual DC input level, and an internal component due to circuit asymmetries.

Based on these design considerations, a novel low-voltage CMOS realization of a precision full-wave rectifier is presented in Figure 3.14, where I_{out} and I_{rect} stand for the incoming output from the Log amplifier and the full-wave rectified signal, respectively. Since current offsets caused by technology mismatching are also related to current bias, the proposed circuit approach minimizes such component by choosing quiescent biasing levels as low as $I_{rectq} \in (1\text{nA}, 10\text{nA})$. However, in order to avoid any incoming offset, a low enough impedance should be also seen at the input to allow DC decoupling as depicted in Figure 3.14. Both design constraints can be achieved by the local feedback loop M1-M8, which ensures a low-enough input impedance even for very low values of current biasing. Two operation cases should be distinguished: while positive phases of I_{out} are processed by M1-M5 devices, negative swings are rectified through M6-M8, so the full rectified waveform I_{rect} is finally collected via M9-M10. Some frequency compensation (C_{comp}) may be required depending on the $I_{rectbias}/I_{rectq}$ ratio.

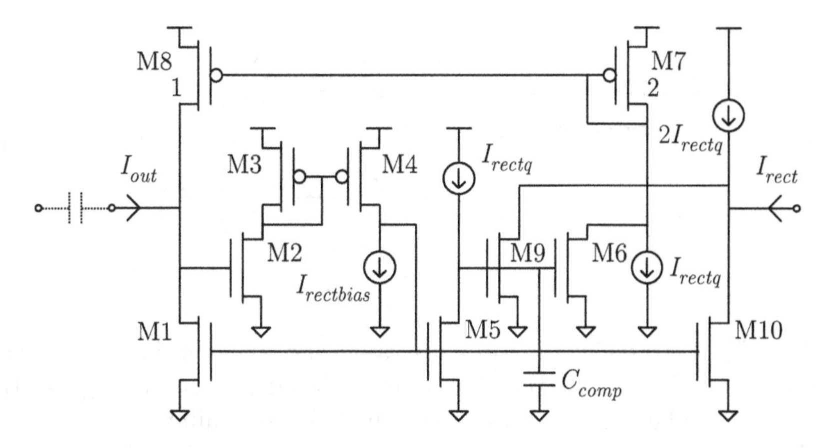

Figure 3.14. Low-voltage CMOS proposal for precision full-wave rectification.

3.4 Envelope Filtering

Typically, frequency selective stages are needed in AGC systems to compute the approximate output envelope ($\tilde{I}_{out}$) from its full-wave rectified representation (I_{rect}). As already pointed out in Section 2, this step usually involves a combination of different first or second order low-pass filters, depending on the desired performance in terms of signal distortion. In any case, the general design problem to be faced can be formulated as finding a solution for the following ordinary differential equation:

$$\frac{\mathrm{d}\tilde{I}_{out}}{\mathrm{d}t} = 2\pi f_o \left(I_{rect} - \tilde{I}_{out} \right) \tag{3.24}$$

where f_o stands for the Corner frequency at -3dB. Arbitrary filtering can be easily synthesized in the Log domain context of this work through the novel low-voltage CMOS circuit techniques presented in Chapter 4 (Please refer to Chapter 4 for a more detailed analysis of the basic building block proposed next).

In the particular case of a low-pass response, the most suitable choice are the non-saturated cells, which are based on the SD compression law of (3.6). From equation (3.24), the required processing in the compressed V-domain can be expressed as:

$$\frac{\mathrm{d}\tilde{V}_{out}}{\mathrm{d}t} = 2\pi f_o U_t \left(1 - e^{\frac{\tilde{V}_{out} - V_{rect}}{U_t}} \right) \tag{3.25}$$

In order to finally obtain a circuit implementation, the internal state-space variable $\tilde{V}_{out}$ is stored across a grounded linear capacitor C. Hence, (3.25) is finally translated into the charge domain (Q) as:

$$\frac{d\tilde{Q}_{out}}{dt} = \underbrace{C\frac{d\tilde{V}_{out}}{dt}}_{I_{cap}} = I_{tuno}\left(1 - e^{\frac{\tilde{V}_{out}-V_{rect}}{U_t}}\right)$$
$$= I_{tuno}e^{\frac{\tilde{V}_{out}}{U_t}}\left(e^{-\frac{\tilde{V}_{out}}{U_t}} - e^{-\frac{V_{rect}}{U_t}}\right) \tag{3.26}$$

$$I_{tuno} = 2\pi f_o U_t C \tag{3.27}$$

where I_{tuno} stands for the tuning parameter of the non-linear transconductance, which controls the charge and discharge current (I_{cap}) of the capacitor C. The splitting in expression (3.26) is similar in form to the MOSFET drain current expression in weak inversion conduction from Table 2.1. In fact, the remaining part of equation (3.26) is equivalent to a signal-dependent gate tuning. Such control can be supplied by a matched device operating in weak inversion saturation, fed at I_{tuno} and sharing the same source bias.

Based on this idea, a low-voltage CMOS implementation of the low-pass envelope filter is proposed in Figure 3.15, where all boxed devices are supposed to operate in weak inversion. The input compression and output expansion processes are performed by M1 and M2 transistors, respectively. The role of the shifter M4-M5 is to optimize automatically the low-voltage operation of all telescopic devices according to the maximum input signal level (I_{max}). The core of the low-pass filter consists of the non-linear transconductance pairs M6-M9 and the voltage followers (VF), while I_{tuno} sources are devoted to f_o tuning according to expression (3.27) (i.e. about 0.16nA/pFKHz at room temperature). The VF block can be implemented using the mismatching-optimized follower already proposed in Figure 3.11 (even with its M2-I_{source} devices removed).

The topology of Figure 3.15 also includes an additional filter ($\tilde{I}_{out2}$) sharing the compressor and a set of MOS switches to select between the first and second order response via a by-pass of the first stage. Based on this approach, a dual first-order envelope filter example for syllabic AGC is presented in Section 4.

3.5 Log Ruler

In the general AGC model of Figure 3.1, the effective output envelope ($\tilde{I}_{out}$) must be translated back to a Log scale for the synthesis of the CR parameter. In this sense, the reference value of such a Log rule is the processed threshold knee (GI_{tk}) itself. Due to the SC strategy chosen in the Log amplifier of Figure 3.8, the SD compression law (3.6) should be

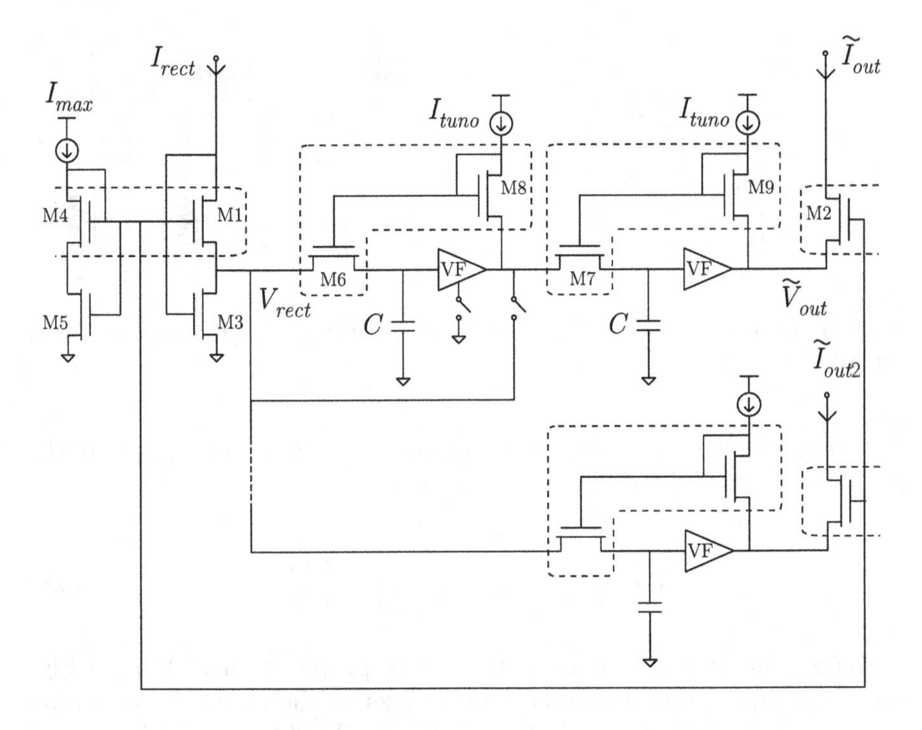

Figure 3.15. Low-voltage CMOS proposal for envelope filtering.

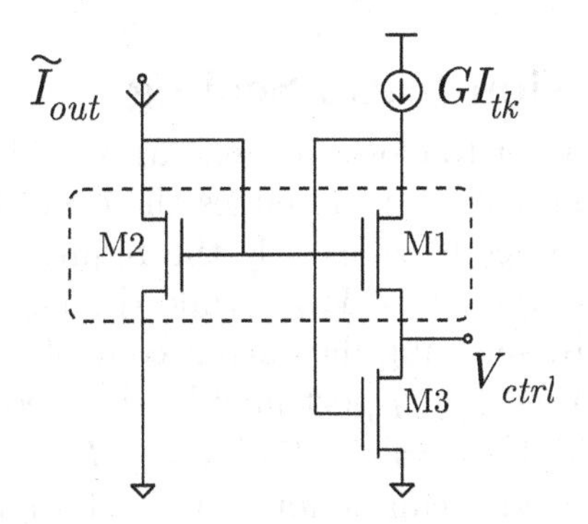

Figure 3.16. Low-voltage CMOS proposal for the Log ruler.

implemented in the logarithmic evaluation of $\tilde{I}_{out}$, so that the resulting control signal (V_{ctrl}) is fully compatible with $V_{gaini,o}$. Based on these specifications, a low-voltage CMOS proposal is shown in Figure 3.16 where the boxed devices are supposed to operate in weak inversion saturation.

In fact, the proposed circuit topology can be understood as the basic GD-SC amplifier of Figure 3.3 with its input and output fed at GI_{tk} and

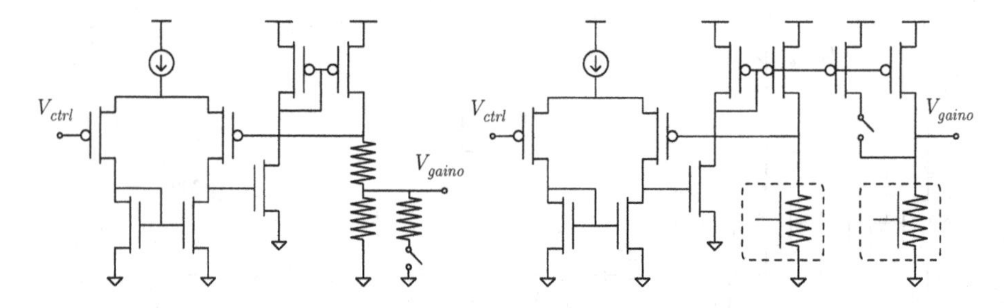

Figure 3.17. Low-voltage CMOS programming of m factor based on resistors (left) and MRCs (right).

$\tilde{I}_{out}$, respectively. According to design equation (3.9) and $V_{gaini} \equiv 0$, the resulting control signal $V_{ctrl} = V_{gaino}$ follows:

$$V_{ctrl} = \begin{cases} 0 & \tilde{I}_{in} \leq I_{tk} \\ U_t \ln\left(\frac{\tilde{I}_{in}}{I_{tk}}\right) & \tilde{I}_{in} > I_{tk} \end{cases} \quad (3.28)$$

Hence, the required first-quadrant only propagation of V_{ctrl} in Figure 3.1 is already implemented here thanks to the boundary condition $V_{ctrl} > 0$. Continuous tuning of the AGC threshold knee point is reduced to programing the I_{tk} current source.

3.6 Compression Ratio Scaling

The final step in order to close the general AGC loop of Figure 3.1 is the linear scaling factor m, which defines the overall compression ratio according to (3.5). Since $0 < m < 1$, the required processing can be seen as computing a fraction of V_{ctrl}, while the negative sign in $-m$ is easily implemented by selecting the output control port $V_{gaino} = mV_{ctrl}$. In that case, the unused V_{gaini} port may be devoted to programing the open loop gain in the linear range (i.e. $\tilde{I}_{in} \leq I_{tk}$) of the AGC system. An illustration of this idea can be found in Section 4

From the point of view of the circuit, the voltage scaler m has to process only the limited range $0 < V_{ctrl} < 8U_t$. Taking into account this particularity, a low-voltage realization is proposed in Figure 3.17 (left). No absolute accuracy, high linearity or large resistance values are required in practice for the resistive divider due to the ratio-based design and the limited range of V_{ctrl}. As a result, monolithic implementations using local wells can usually meet the relaxed specifications for such components. In any case, an all-MOS approach based on a new grounded MOS resistive circuit (MRC) is also proposed in Figure 3.17 (right) to avoid the use of passive resistors.

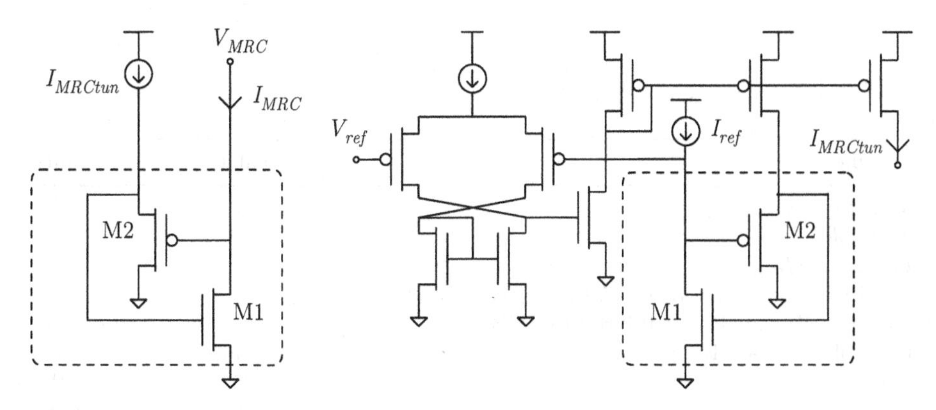

Figure 3.18. Low-voltage CMOS proposal of the grounded MRC (left) and its auto-tuning circuitry (right).

A novel low-voltage implementation of the grounded MRC block is presented in Figure 3.18 (left). Linearization of the equivalent resistance (R_{MRC}) is based on operating the core device M1 in the strong inversion conduction region. The MOS drain current equation in this regime follows from Table 2.1 and $V_{SB} \equiv 0$:

$$I_D = \beta \left[(V_{GB} - V_{TO}) - \frac{n}{2} V_{DB} \right] V_{DB} \tag{3.29}$$

The basic idea is to introduce a voltage compensation at the gate to obtain a linear incremental behaviour in terms of $R_{MRC} = \Delta I_D / \Delta V_{DB}$. From (3.29), the required compensation is found to be ideally $\Delta V_{GB} = \frac{n_N}{2} \Delta V_{DB}$. The feedback device M2 of Figure 3.18 operating in saturation generates a gate voltage compensation $\Delta V_{GB} = \frac{1}{n_P} \Delta V_{DB}$, which is an exact solution for $n_N n_P \equiv 2$. In general, some quadratic term remains in the MRC impedance as:

$$I_{MRC} = \frac{V_{MRC}}{R_{MRC}} + \beta_1 \left(\frac{1}{n_P} - \frac{n_N}{2} \right) V_{MRC}^2 \tag{3.30}$$

where:

$$R_{MRC} = \frac{1}{\beta_1} \left[\frac{n_P - 1}{n_P} V_{DD} - \frac{V_{TOP}}{n_P} - V_{TON} - \sqrt{\frac{2 I_{MRCtun}}{n_P \beta_2}} \right]^{-1} \tag{3.31}$$

In order to simplify the adjustment of R_{MRC} through I_{MRCtun} in (3.31), a self-tuning circuit is also proposed in Figure 3.18 (right), resulting in an effective $R_{MRC} = V_{ref} / I_{ref}$.

Table 3.3. General amplifier Performances.

Parameter	Value	Units
Min. Supply Voltage	1.0	V
Technology $(V_{TON} + \|V_{TOP}\|)_{max}$	1.3	V
Full-Scale (I_{max})	4	μA_{pp}
Total Harmonic Distortion @50%I_{max} G=+30dB	0.6	%
Dynamic Range (100Hz-10KHz) G=0dB	72 – 78	dB
Controllable Gain Range (G)	±40	dB
Power	25	μW
Si Area	0.16	mm^2

Integrable values of resistance typically range $R_{MRC} \in (10K\Omega, 500K\Omega)$, like the experimental example of Figure 3.19. Apart from the calibrating and tuning possibilities, important Si area savings can be obtained with the proposed MRC compared to the equivalent passive implementation through n-well resistors. For example, taking the same CMOS process of Figure 3.19 with $R_{n-well} \simeq 1K\Omega/\square$ and 10μm of minimum width and pitch, the required n-well area would be increased by a factor of 4. Typical results of the novel MOS resistive circuit such as the one in Figure 3.20 exhibit enough linearity, as these blocks are devoted to controlling computation rather than the actual signal processing.

4. Design Examples

Two design examples are presented here as demonstrators of the proposed CMOS circuit techniques. Specifications have been chosen for their usage in very low-voltage portable audio applications like hearing-aids-on-chip. The target technology in both cases is a 1.2μm CMOS double-metal double-poly-Si process. The first circuit is depicted in Figure 3.21 and consists of the general purpose amplifier topology of Figure 3.8 including both gain control voltage sources. The overall performance of this implementation is summarized in Table 4.4. Detailed graphical results can be also seen in Figures 3.22, 3.23 and 3.24[1].

The second design example is a complete AGC stage built around the previous amplifier cell and following the high-level model of Figure 3.1. Its low-voltage CMOS implementation is based on all the basic building blocks proposed in previous sections. In this case, two first-order

[1]Distortion results match with the theoretical expressions of Chapter 7 since $(W/L)_{1,2} = 20 \times 80\mu m/3\mu m$ and $I_{Su} = 135nA$, so $IC_{max} = 0.042$ at 50% of full-scale.

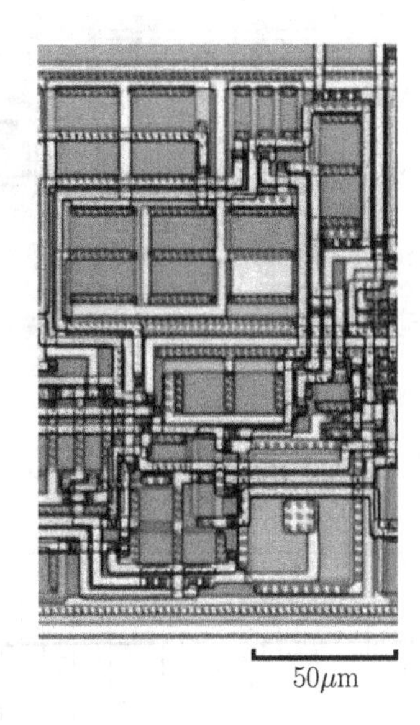

Figure 3.19. Microscope photography of a 3×186KΩ MRC implementation.

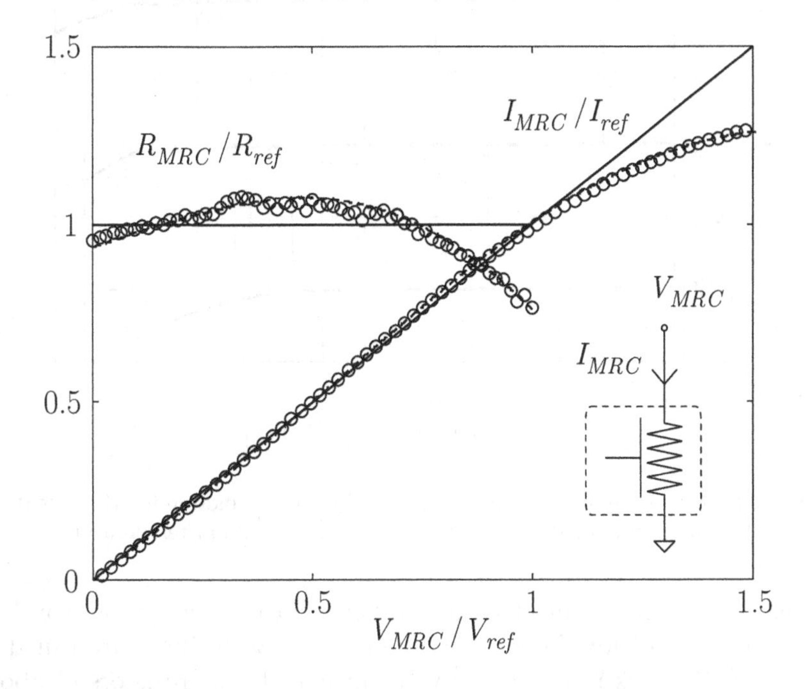

Figure 3.20. Experimental (dotted), simulated (dashed) and ideal (solid) V/I and resistance curves of the proposed grounded MRC for $R_{ref} = V_{ref}/I_{ref} = 186\text{mV}/1\mu\text{A} = 186\text{K}\Omega$.

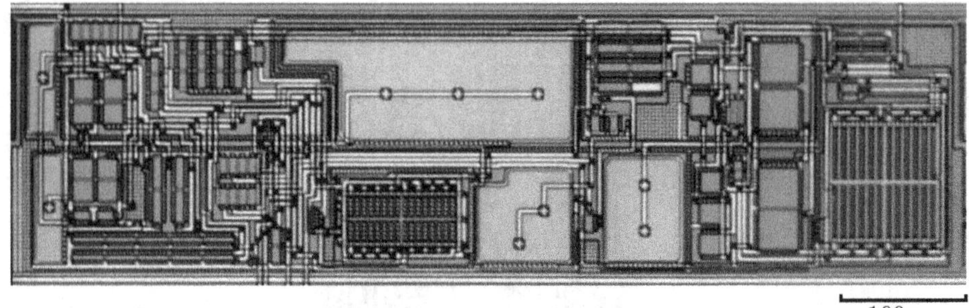

Figure 3.21. Microscope photography of an amplifier example.

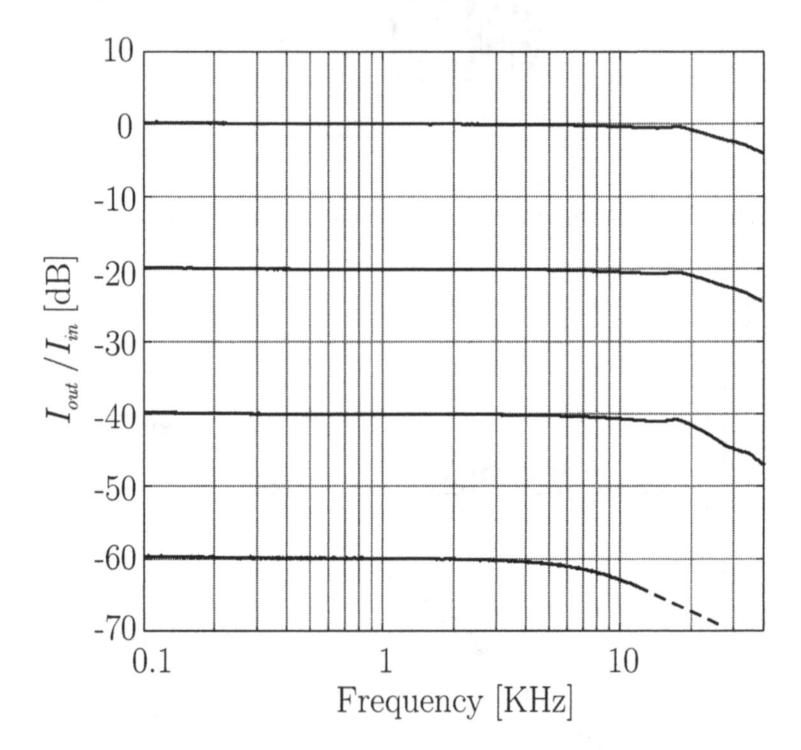

Figure 3.22. Experimental amplifier large signal transfer function for $V_{gaini} \equiv 0$ and $V_{gaino} = 0$ (upper), 62mV, 124mV and 186mV (lower) at room temperature.

low-pass filters are connected in parallel, sharing the compressor M5-M8 of Figure 3.15 for the envelope detection. Both filters are tuned at I_{tuno} = 10nA, thus integrating two equivalent linear resistors of about 2.5MΩ. Corner frequencies of 14Hz and 0.6Hz are selected using external capacitors of 4.7nF and 100nF, respectively. The transient behaviour results in an attack time of t_{att} = 15ms and release times adapted to the

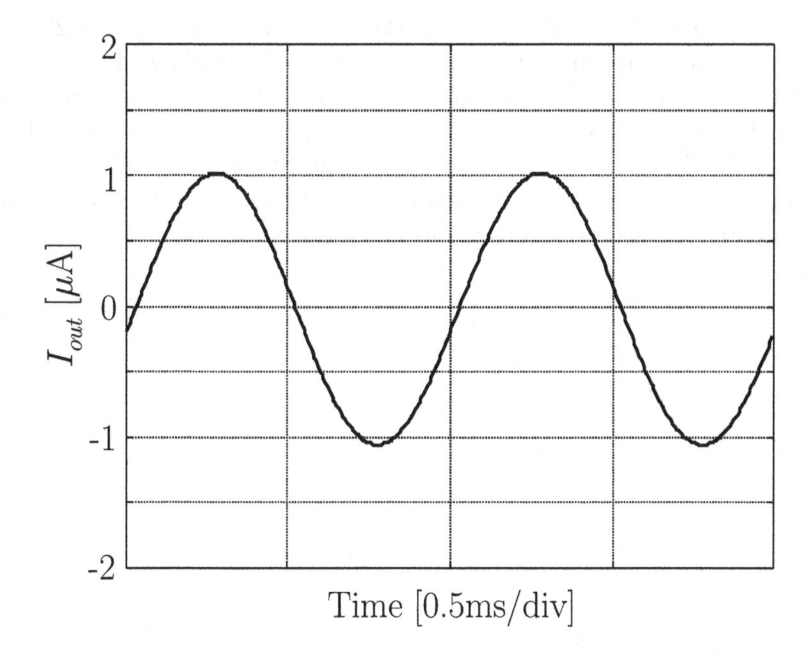

Figure 3.23. Experimental amplifier output at 50% of full-scale.

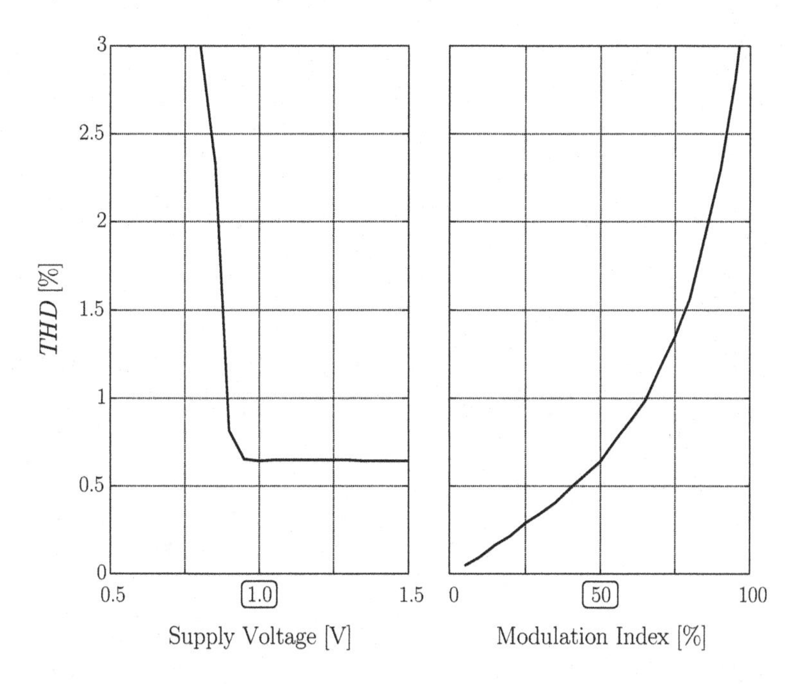

Figure 3.24. Simulated THD of the amplifier cell at $G = +30\text{dB}$. Boxed values indicate default conditions.

input burst duration of up to $t_{rel} < 500$ms, as shown in Figure 3.25. The flexibility of the presented CMOS circuit techniques allows independent programming of the AGC threshold knee point $I_{tk} > 3$nA, the compression ratio $CR \in (1, \infty)$, as well as the open loop gain $G \in (0, +40\text{dB})$ through the V_{gaini} port. All this configurability is summarized in Figure 3.26. The complete AGC stage can truly operate at 1.0V supply, while exhibiting a power consumption as low as 60μW.

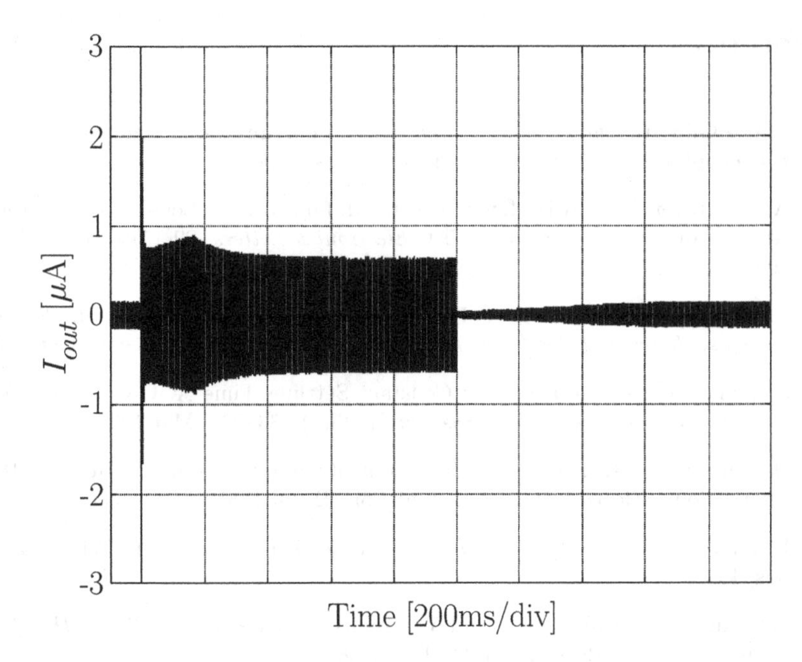

Figure 3.25. Simulated AGC transient output when a $20\text{nA}_{peak} \pm 25\text{dB}$ and 1000ms width input burst is applied with $G = +20\text{dB}$, $I_{tk} = 10\text{nA}_{rms}$ and $CR = 2:1$.

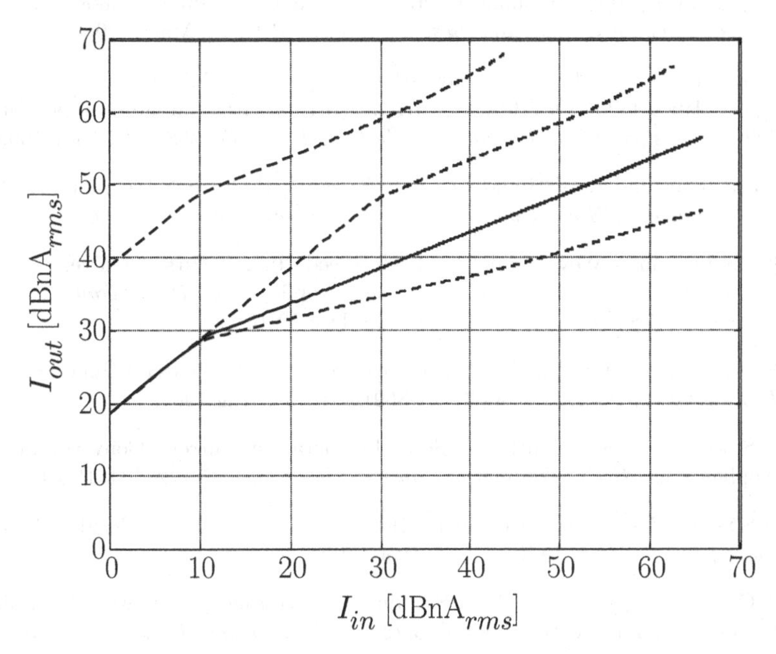

Figure 3.26. Simulated AGC steady-state response for $G = +20\text{dB}$, $CR = 2:1$ and $I_{tk} = 3\text{nA}_{rms}$ (solid). Tuning capabilities are shown in dashed lines for $G = +40\text{dB}$ (upper), $I_{tk} = 30\text{nA}_{rms}$ (middle) and $CR = 3:1$ (lower).

References

[1] Barrie Gilbert. *Current-mode Circuits From a Translinear Viewpoint: A Tutorial*, chapter 2, pages 11–92. Peter Peregrinus, 1990.

[2] A. van Staveren and A.H.M. van Roermund. Low-Voltage Low-Power Controlled Attenuator for Hearing Aids. *IEE Electronics Letters*, 29(15):1355–1356, Jul 1993.

[3] F.Floru. Attack and Release Time Constants in RMS-Based Feedback Compressors. *Journal of the Audio Engineering Society*, 47(10):788–804, Oct 1999.

[4] J.M.Khoury. On the Design of Constant Settling Time AGC Circuits. *IEEE Transactions on Circuits and Systems-II*, 45(3):283–294, Mar 1998.

[5] Hearing Aids with Automatic Gain Control Circuits. Technical Report 118-2, International Electrotechnical Commission (IEC), 1983.

[6] E.A.Vitoz. *Dynamic Current Mirrors*, chapter 7, pages 297–326. Peter Peregrinus, 1990.

[7] S.J.Daubert, D.Vallancourt, and Y.Tsividis. Current Copier Cells. *IEE Electronics Letters*, 24(25):1560–1562, Dec 1988.

[8] D.G.Nairn and C.A.T.Salama. A Ratio-Independent Algorithmic Analog-to-Digital Converter Combining Current Mode and Dynamic Techniques. *IEEE Transactions on Circuits and Systems-I*, 37(3):319–325, Mar 1990.

[9] T.Serrano-Gotarredona, B.Linares-Barranco, and A.G.Andreou. Very Wide Range Tunable CMOS/Bipolar Current Mirrors with Voltage Clamped Input. *IEEE Transactions on Circuits and Systems-I*, 46(11):1398–1407, Nov 1999.

[10] J.B.Hughes and K.W.Moulding. S^2I: A Switched-Current Technique for High Performance. *IEE Electronics Letters*, 29(16):1400–1401, Aug 1993.

[11] R.Huang and C.Wey. Design of High-Speed High-Accuracy Current Copiers for Low-Voltage Analog Signal Processing Applications. *IEEE Transactions on Circuits and Systems-II*, 43(12):836–839, Dec 1996.

[12] K.C.Smith and A.Sedra. The Current Conveyor - A New Circuit Building Block. *Proceedings of the IEEE*, 56(1368–1369):353–365, Aug 1968.

[13] A.S.Sedra and K.C.Smith. A Second-Generation Current Conveyor and Its Applications. *IEEE Transactions on Circuit Theory*, 17:132–134, Feb 1970.

[14] A.S.Sedra. The Current Conveyor: History, Progress and New Results. *Proceedings of the IEEE*, 137-G(2):78–87, Apr 1990.

[15] A.C. van der Woerd and W.A.Serdijn. Low-Voltage Low-Power Controllable Preamplifier for Electret Microphones. *IEEE Journal of Solid State Circuits*, 29(9):1052–1055, Oct 1993.

Chapter 4

FILTERING

Abstract The following chapter includes all the new circuit techniques related to frequency selective stages. After reviewing the Log companding principle of operation, the generalization for the MOS transistor is presented. Based on these results, different types of CMOS basic building blocks for arbitrary Log filters are proposed in conjunction with a compact synthesis methodology. This research is extended to all-MOS implementations for their integration through digital CMOS technologies. Finally, some filter cases and design examples are presented as demonstrators.

1. Log Companding Principle

Due to the wide variety of frequency selective transfer functions, this introduction will only focus on the integrator as the basic example to illustrate the Log companding principle of signal filtering. Referred to the normalized nomenclature introduced in Chapter 1, the integrator output satisfies:

$$\frac{\mathrm{d}y_{out}}{\mathrm{d}t} \doteq \frac{y_{in}}{\tau} \tag{4.1}$$

with τ being the general time constant. The above ordinary differential equation must now be translated to the compressed x-domain in order to study the equivalent non-linear internal processing which preserves the external linearity of the system. As first proposed in [1] using diodes, this step is performed by applying the Chain Rule to the previous expression, using the Log companding function F defined in (1.5):

$$\frac{\mathrm{d}y_{out}}{\mathrm{d}t} = \frac{\mathrm{d}y_{out}}{\mathrm{d}x_{out}}\frac{\mathrm{d}x_{out}}{\mathrm{d}t} \equiv y_{out}\frac{\mathrm{d}x_{out}}{\mathrm{d}t} \tag{4.2}$$

It is clear at this point that one of the advantages of Log F functions is the simple equivalence between y and x domain derivatives. This special property is common to all translinear devices, for which its transconductance ($\mathrm{d}F/\mathrm{d}x$) shows a linear dependence with respect to its current (y), as with exponential and hyperbolic F laws.

Thus, in order to integrate the necessary processing physically in the x-domain, x_{out} is linearly controlled through the voltage (x_{cap}) stored across a linear capacitor (C):

$$x_{out} \doteq a x_{cap} + b \tag{4.3}$$

According to the above linear relation, the derivative has a direct correspondence in the compressed domain:

$$\frac{\mathrm{d}y_{out}}{\mathrm{d}t} = y_{out}\frac{\mathrm{d}x_{out}}{\mathrm{d}x_{cap}}\frac{\mathrm{d}x_{cap}}{\mathrm{d}t} \equiv y_{out}\frac{\mathrm{d}x_{cap}}{\mathrm{d}t} \tag{4.4}$$

where $a \equiv 1$ and $b \equiv 0$ has been chosen for simplification. As proposed in [2], the necessity of inductive components from (4.1) can be skipped in favor of integrable capacitive elements by the use of a tuning current (y_{tun}):

$$y_{out}\underbrace{C\frac{\mathrm{d}x_{cap}}{\mathrm{d}t}\frac{U_t}{I_S}}_{y_{cap}} = y_{tun}y_{in} \qquad y_{tun} \doteq \frac{CU_t}{\tau I_s} \tag{4.5}$$

so that the underbraced term corresponds to the charging and discharging current (y_{cap}) of C. As a result, the above equation allows an equivalent expression in terms of a product of currents:

$$y_{out}y_{cap} = y_{tun}y_{in} \tag{4.6}$$

Such a product can be synthesized through the Translinear Principle (TP) [3, 4]. Unfortunately, the TP was originally intended for bipolar elements (i.e. diodes and BJTs) which present first-order differences with respect to the MOS transistor as already explained in Chapter 1. In consequence, the next section presents a new TP rewritten for the MOSFET as a previous step before to introducing the CMOS basic building blocks for Log filtering.

2. CMOS Generalization

The well-known Translinear Principle was originally oriented to bipolar semiconductor devices, such as diodes and BJTs operating in their active region. Hence, all device currents were supposed to be controlled

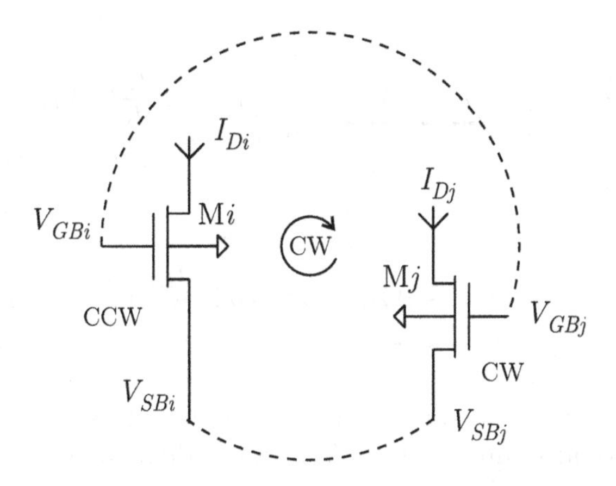

Figure 4.1. General NMOS translinear loop.

by differential voltages. Although the MOSFET biased at weak inversion saturation exhibits a similar exponential function from Table 2.1, its drain current asymmetry from gate to source or drain voltages caused by non-unitary subthreshold slopes ($n \neq 1$) and non-local substrates ($V_{SB,DB} \neq 0$), requires a new evaluation of the TP validity for such devices. The following study presents the topological restrictions that any CMOS basic building block must verify to implement products based on the TP, as in (4.6).

The starting structure is a Translinear Loop (TL) made of MOS devices operating in weak inversion saturation and connected through gate and source as depicted in Figure 4.1 (results can be easily generalized to either reverse saturation and PMOS transistor types).

Due to the anti-latch-up rules of CMOS technologies, no bulk-driven topologies will be considered here as already argued in Chapter 3. Thus, all local bulks are always connected to voltage supplies (V_{SS} and V_{DD} for NMOS and PMOS transistors, respectively) unless specified. Following the TP approach, the main equation is first obtained by applying the KVL to the loop:

$$\sum_{\text{CCW}} (V_{GB} - V_{SB})_i = \sum_{\text{CW}} (V_{GB} - V_{SB})_j \tag{4.7}$$

where CCW and CW stand for counter-clock and clockwise, respectively, taking the gate-to-source reference. In order to obtain the final product of currents, the above equation must be rewritten in terms of a sum of logs as:

$$\sum_{\text{CCW}} n_i U_t \ln \frac{I_{Di}}{I_{Si}} + V_{TOi} + \underbrace{(n_i - 1)V_{SBi}} = \sum_{\text{CW}} n_j U_t \ln \frac{I_{Dj}}{I_{Sj}} + V_{TOj} + \underbrace{(n_j - 1)V_{SBj}} \quad (4.8)$$

or:

$$\sum_{\text{CCW}} U_t \ln \frac{I_{Di}}{I_{Si}} + \frac{V_{TOi}}{n_i} + \underbrace{(1 - \frac{1}{n_i})V_{GBi}} = \sum_{\text{CW}} U_t \ln \frac{I_{Dj}}{I_{Sj}} + \frac{V_{TOj}}{n_j} + \underbrace{(1 - \frac{1}{n_j})V_{GBj}} \quad (4.9)$$

However, even supposing global thermal and technological matching, which would eliminate U_t, V_{TO} and n dependencies, some signal-dependent terms marked in both expressions still remain, causing distortion.

The new solution proposed here solves the above problem by building all MOS TLs by means of matched pairs in gate-driven (GD) (i.e. common source and bulk $V_{SBi} \equiv V_{SBj}$) or source-driven (SD) (i.e. common gate and bulk $V_{GBi} \equiv V_{GBj}$) configurations. Such a rearrangement enables one-by-one cancellation of all signal-dependent marked terms in (4.8) and (4.9), thus obtaining the desired product of currents:

$$\prod_{\text{CCW}} \left(\frac{I_D}{I_S}\right)_i = \prod_{\text{CW}} \left(\frac{I_D}{I_S}\right)_j \quad (4.10)$$

As a result, MOS loops must always be built by means of the allowed topologies listed in Figure 4.2.

Furthermore, only a single cell type from the above table should be used in the same TL. This requirement of not mixing GD and SD elements can be clearly seen in the example of Figure 4.3. The general expression (4.7) is particularized in this case to:

$$(V_1 - V_3) + (V_3 - V_4) = (V_1 - V_2) + (V_2 - V_4) \quad (4.11)$$

and developed as:

$$U_t \ln \left(\frac{I_D}{I_S}\right)_1 + \frac{V_{TO1}}{n_1} + (1 - \frac{1}{n_1})V_1 + n_3 U_t \ln \left(\frac{I_D}{I_S}\right)_3 + V_{TO3} + (n_3 - 1)V_4 =$$
$$U_t \ln \left(\frac{I_D}{I_S}\right)_2 + \frac{V_{TO2}}{n_2} + (1 - \frac{1}{n_2})V_1 + n_4 U_t \ln \left(\frac{I_D}{I_S}\right)_4 + V_{TO4} + (n_4 - 1)V_4 \quad (4.12)$$

As argued, matched pairs enable cancellation of V_{TO} and signal-dependent terms V_1 and V_4:

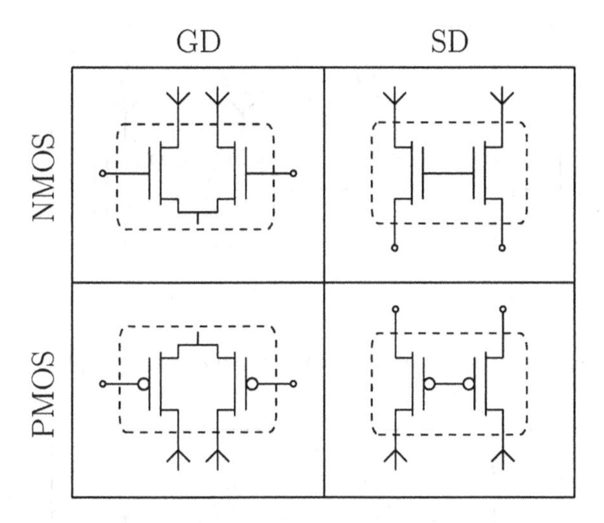

Figure 4.2. Proposed MOS TL elements. Matched devices and $V_B = V_{SS/DD}$ for NMOS/PMOS are supposed.

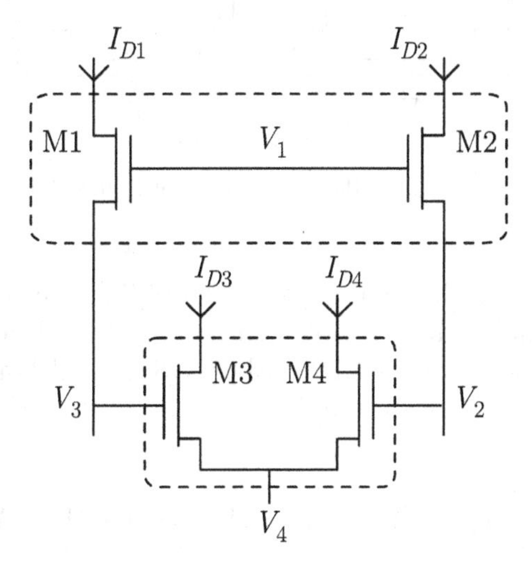

Figure 4.3. Example of mixed driven NMOS TL.

$$U_t \ln\left(\frac{I_D}{I_S}\right)_1 + n_3 U_t \ln\left(\frac{I_D}{I_S}\right)_3 = U_t \ln\left(\frac{I_D}{I_S}\right)_2 + n_4 U_t \ln\left(\frac{I_D}{I_S}\right)_4 \quad (4.13)$$

However, even overall isothermal operation does not return the final desired product, but:

$$\left(\frac{I_D}{I_S}\right)_1 \left(\frac{I_D}{I_S}\right)_3^{n_3} = \left(\frac{I_D}{I_S}\right)_2 \left(\frac{I_D}{I_S}\right)_4^{n_4} \quad (4.14)$$

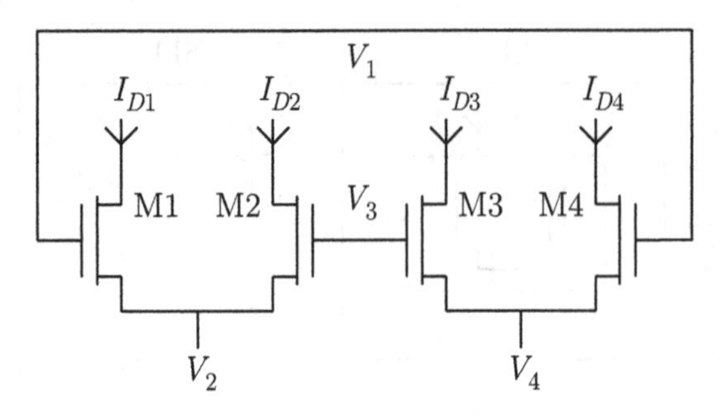

Figure 4.4. Both GD and SD MOS TL example.

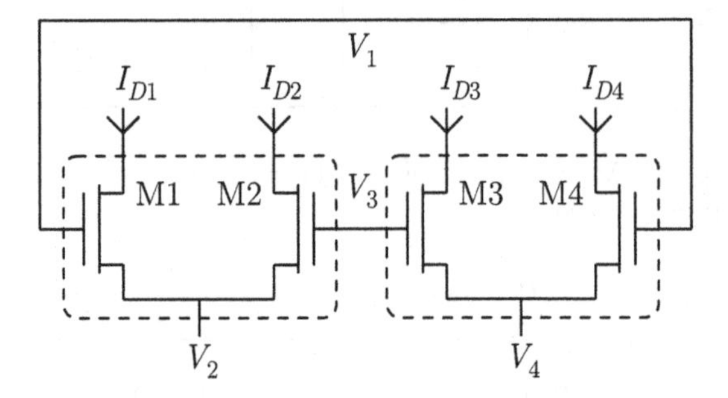

Figure 4.5. GD arrangement of Figure 4.4.

Consequently, two different types of MOS TLs can be generally distinguished. Consider the example of Figure 4.4, with a particular loop which can be understood as GD or SD MOS TL. Although the starting and final design equations are unique,

$$(V_1 - V_2) + (V_3 - V_4) = (V_3 - V_2) + (V_1 - V_4) \tag{4.15}$$

$$\left(\frac{I_D}{I_S}\right)_1 \left(\frac{I_D}{I_S}\right)_3 = \left(\frac{I_D}{I_S}\right)_2 \left(\frac{I_D}{I_S}\right)_4 \tag{4.16}$$

the mismatch requirements differ depending on the basic cell type:

GD. Transistors are arranged in gate-driven matched pairs named M1-M2 and M3-M4 as indicated in Figure 4.5, developing (4.15) into:

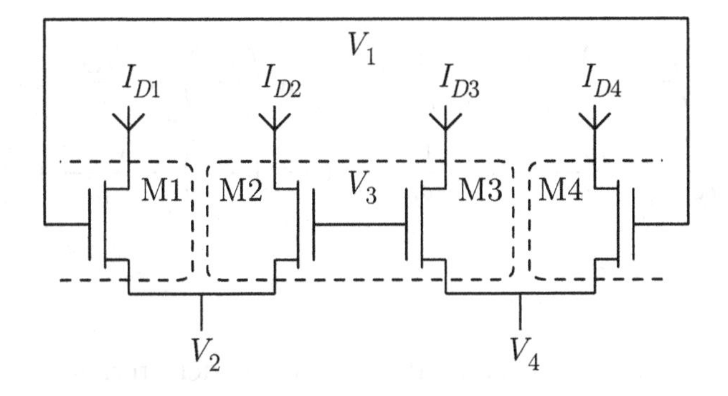

Figure 4.6. SD arrangement of Figure 4.4.

$$n_1 U_t \ln\left(\frac{I_D}{I_S}\right)_1 + V_{TO1} + (n_1 - 1)V_2 + n_3 U_t \ln\left(\frac{I_D}{I_S}\right)_3 + V_{TO3} + (n_3 - 1)V_4 =$$
$$n_2 U_t \ln\left(\frac{I_D}{I_S}\right)_2 + V_{TO2} + (n_2 - 1)V_2 + n_4 U_t \ln\left(\frac{I_D}{I_S}\right)_4 + V_{TO4} + (n_4 - 1)V_4 \quad (4.17)$$

The signal-dependent parts can be canceled in:

$$n_1 U_t \ln\left(\frac{I_D}{I_S}\right)_1 + n_3 U_t \ln\left(\frac{I_D}{I_S}\right)_3 = n_2 U_t \ln\left(\frac{I_D}{I_S}\right)_2 + n_4 U_t \ln\left(\frac{I_D}{I_S}\right)_4 \quad (4.18)$$

However, some extra technological parameters still remain in the isothermal product:

$$\left(\frac{I_D}{I_S}\right)_1^{n_1} \left(\frac{I_D}{I_S}\right)_3^{n_3} = \left(\frac{I_D}{I_S}\right)_2^{n_2} \left(\frac{I_D}{I_S}\right)_4^{n_4} \quad (4.19)$$

Hence, GD MOS TLs require physical matching of all devices in the loop (i.e. M1 to M4) to obtain the desired final design equation (4.16).

SD. The same MOS TL, however, can be rearranged using the source-driven pair distribution M2-M3 and M1-M4 of Figure 4.6. Now, according to the new matched groups, (4.15) is rewritten as:

$$U_t \ln\left(\frac{I_D}{I_S}\right)_1 + \frac{V_{TO1}}{n_1} + (1-\frac{1}{n_1})V_1 + U_t \ln\left(\frac{I_D}{I_S}\right)_3 + \frac{V_{TO3}}{n_3} + (1-\frac{1}{n_3})V_3 =$$
$$U_t \ln\left(\frac{I_D}{I_S}\right)_2 + \frac{V_{TO2}}{n_2} + (1-\frac{1}{n_2})V_3 + U_t \ln\left(\frac{I_D}{I_S}\right)_4 + \frac{V_{TO4}}{n_4} + (1-\frac{1}{n_4})V_1 \tag{4.20}$$

In this case, after canceling all signal-dependent terms:

$$U_t \ln\left(\frac{I_D}{I_S}\right)_1 + U_t \ln\left(\frac{I_D}{I_S}\right)_3 = U_t \ln\left(\frac{I_D}{I_S}\right)_2 + U_t \ln\left(\frac{I_D}{I_S}\right)_4 \tag{4.21}$$

no extra technological parameters appear in the product expression and only an isothermal operation is necessary. Hence, SD MOS TLs only require technological matching at the pair level (i.e. M2-M3 and M1-M4).

In conclusion, GD can be understood as a subcategory of SD MOS TLs, as declared in the following generalized Translinear Principle for MOS devices of Figure 4.7 (compatible results are reported by other similar theoretical analysis [5, 6, 7]).

3. Basic Building Blocks

General filter specifications are usually expressed in terms of an arbitrary frequency selective transfer function in the s-domain and order N, as in the following expression for the one-input one-output case:

$$H(s) = \frac{a_N s^N + a_{N-1} s^{N-1} + \cdots + a_1 s + a_0}{b_N s^N + b_{N-1} s^{N-1} + \cdots + b_1 s + b_0} \tag{4.22}$$

The resulting structures usually require multiple MOS TLs, which are more difficult to identify than those in Figure 4.1. Hence, the theoretical shortcut proposed by [8] will be selected here to rewrite the product of (4.6) in terms of direct x-domain signals:

$$\underbrace{C\frac{\mathrm{d}x_{cap}}{\mathrm{d}t}}_{y_{cap}} = y_{tun} e^{x_{in} - x_{cap}} \tag{4.23}$$

From the point of view of the circuit, the above expression defines the internal non-linear transconductance required to achieve external linearity, as symbolized in Figure 4.10.

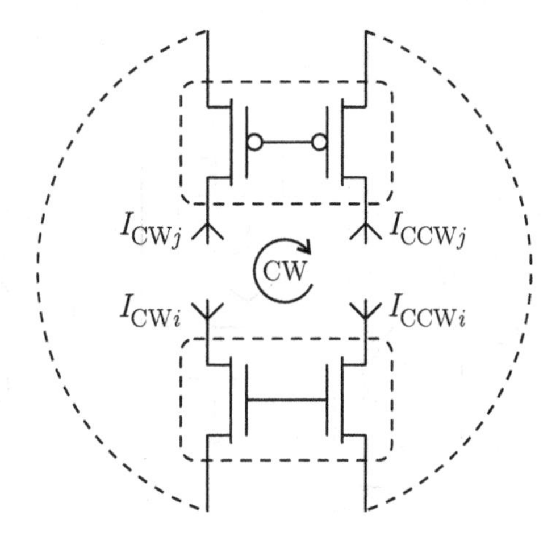

In an isothermal closed loop containing an even number of saturated weak inverted MOS devices with non-isolated bulks, arranged so that they are grouped in common gate matched pairs, the product of the drain current densities in the clock-wise direction is equal to the product of the drain current densities in the counterclockwise direction.

$$\prod_{\text{CW}} \left(\frac{I_D}{I_S}\right) = \prod_{\text{CCW}} \left(\frac{I_D}{I_S}\right)$$

Figure 4.7. Rewriting the Translinear Principle for MOS devices.

In order to generalize results, an equivalent matrix description based on the state-space (SS) representation [9] will be used for (4.22):

$$\begin{cases} \dfrac{\mathrm{d}I}{\mathrm{d}t} = AI + BI_{in} \\ \\ I_{out} = CI + DI_{in} \end{cases} \tag{4.24}$$

Careful attention must be paid when selecting the filter representation in the I-domain since all linear variables must verify $I \in \mathbb{R}^+ \ \forall t$ to allow their internal Log mapping. This issue is addressed in the design methodology proposed in Section 4. In any case, the second SS equation in (4.24) can be easily synthesized in the linear current domain through simple KCL algebra. Thus, the major design problem to be faced is reduced to implementing N first-order ordinary differential equations symbolized for the i-case as:

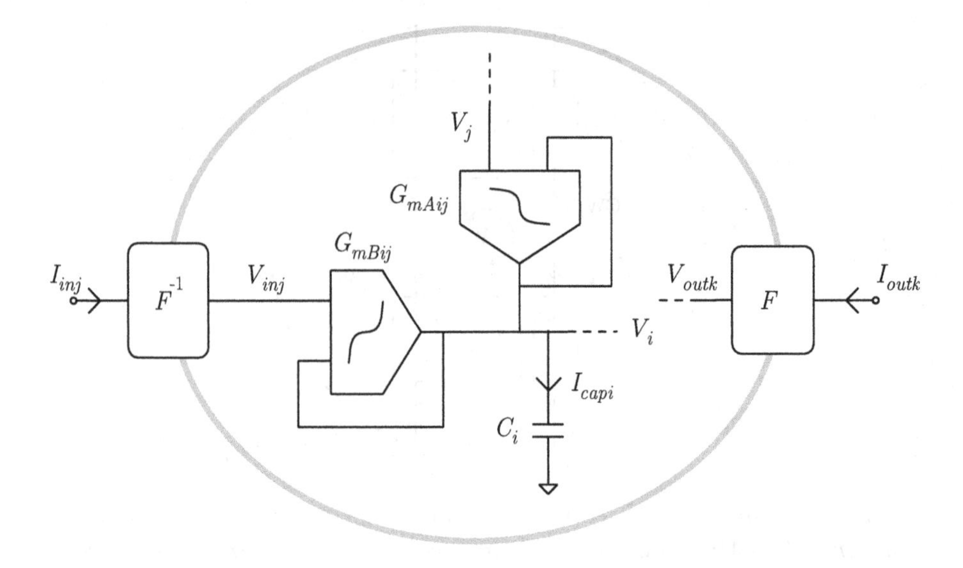

Figure 4.8. General transconductance synthesis approach.

$$\frac{\mathrm{d}I_i}{\mathrm{d}t} = \sum_{j=1}^{N} A_{ij} I_j + \sum_{j=1}^{M} B_{ij} I_{inj} \qquad \text{for } i\text{=1 to } N \tag{4.25}$$

The Chain Rule strategy is now applied to each row in order to translate specifications into the equivalent SS V-domain. At this point and as already mentioned in Chapter 1, circuit techniques are specific for the particular Log companding function F chosen. The next subsections propose different CMOS possibilities, all of them compatible with Figure 4.2.

3.1 Saturated CMOS Cells

As introduced in Chapter 3, three Log companding F functions can be obtained at the transistor level from a MOSFET operating in weak inversion saturation according to the terminal used to compress the internal voltage signal: gate- (GD), bulk- (BD) or source-driven (SD), as illustrated in Figure 4.9.

Following the device model equations from Table 2.1,

$$I = F(V) = \begin{cases} I_S e^{-\frac{V_{TO}+nV_{bias}}{nU_t}} e^{\frac{V}{nU_t}} & \text{GD} \\ I_S e^{\frac{V_{bias}-V_{TO}}{nU_t}} e^{\left(1-\frac{1}{n}\right)\frac{V}{U_t}} & \text{BD} \\ I_S e^{\frac{V_{bias}-V_{TO}}{nU_t}} e^{-\frac{V}{U_t}} & \text{SD} \end{cases} \tag{4.26}$$

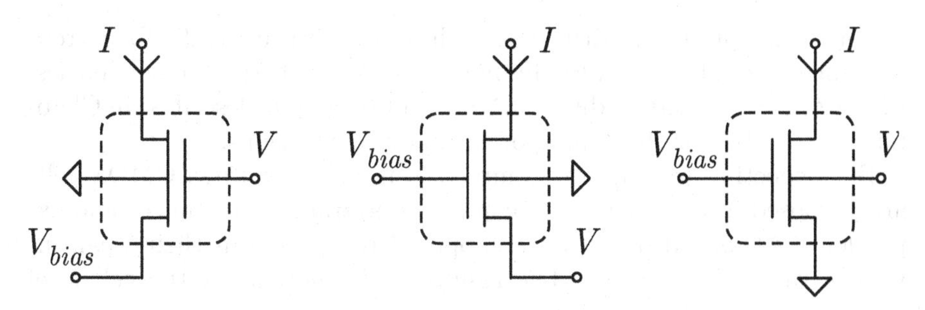

Figure 4.9. Transistor level GD (left), SD (center), and BD (right) implementation of the $F(x) = e^x$ Log Companding function (auxiliary circuitry not shown for simplification).

the following study is limited to GD and SD F functions only. No bulk-driven strategies [10, 11, 12] will be considered in this work due to their compatibility problems with anti-latch-up rules of CMOS technologies. Applying (4.26) to (4.25), the equivalent non-linear processing in the compressed V-domain is found to be:

$$\frac{\mathrm{d}V_i}{\mathrm{d}t} = \begin{cases} \sum_{j=1}^{N} A_{ij} n U_t e^{\frac{V_j - V_i}{nU_t}} + \sum_{j=1}^{M} B_{ij} n U_t e^{\frac{V_{inj} - V_i}{nU_t}} & \text{GD} \\ -\sum_{j=1}^{N} A_{ij} U_t e^{\frac{V_i - V_j}{U_t}} - \sum_{j=1}^{M} B_{ij} U_t e^{\frac{V_i - V_{inj}}{U_t}} & \text{SD} \end{cases} \tag{4.27}$$

From the viewpoint of the circuit, the final non-linear differential equation to be implemented in the charge domain (Q) results in the following transconductances:

$$\frac{\mathrm{d}Q_i}{\mathrm{d}t} = \underbrace{C_i \frac{\mathrm{d}V_i}{\mathrm{d}t}}_{I_{capi}} = \begin{cases} \sum_{j=1}^{N} I_{tunAij} e^{\frac{V_j - V_i}{nU_t}} + \sum_{j=1}^{M} I_{tunBij} e^{\frac{V_{inj} - V_i}{nU_t}} & \text{GD} \\ -\sum_{j=1}^{N} I_{tunAij} e^{\frac{V_i - V_j}{U_t}} - \sum_{j=1}^{M} I_{tunBij} e^{\frac{V_i - V_{inj}}{U_t}} & \text{SD} \end{cases} \tag{4.28}$$

where I_{capi} identifies the capacitor current, and tuning parameters are defined in the A_{ij} case as:

$$I_{tunAij} = \begin{cases} nU_t C_i A_{ij} & \text{GD} \\ U_t C_i A_{ij} & \text{SD} \end{cases} \tag{4.29}$$

For both options, positive-to-absolute-temperature (PTAT) current references should be used to eliminate first-order thermal dependencies. In this sense, the companding PTAT circuit technique described in Chapter 5 is suitable for the tuning of the above expressions.

By inspecting (4.28), two complete sets of low-voltage CMOS cells are proposed in Figure 4.10 for compressors, transconductances and expanders. All boxed devices are supposed to operate in their weak inversion saturation region. The resulting F functions at the cell level are:

$$I = F(V) = \begin{cases} I_{bias} e^{\frac{V - V_{ref}}{nU_t}} & \text{GD} \\ I_{bias} e^{\frac{V_{ref} - V}{U_t}} & \text{SD} \end{cases} \tag{4.30}$$

where I_{bias} and V_{ref} are reference values for the linear and compressed domains, respectively. The separated definition of compressor and expander basic building blocks allows independence between signal range (I_{bias}) and frequency tuning (I_{tun}) specifications. Also, such splitting simplifies multi-path propagation of compressed input signals. Furthermore, all resulting filter topologies are compatible with syllabic circuit techniques such as [20].

Notice that transconductances for A and B coefficients are synthesized to only charge or discharge the SS variable stored in C_i. Such simplification is derived from the Log nature of I/V compression which requires $I > 0$ $\forall t$, otherwise $V \rightarrow -\infty$ and $+\infty$ for the GD and SD cases, respectively. This idea can be seen more clearly in the graphical representation of the non-linear transconductances required for each coefficient in Figure 4.14. In practice, operation point stability is ensured at the SS matrix level as addressed in Section 4.

The role of the operational transresistance amplifier (ORA) is to supply a low enough input impedance at the compressor transistor to minimize the channel length modulation effects, as well as to enable optional linear V/I conversion through a simple input resistor. When not required, this block can be directly replaced by a short-circuit. Also in the SD case, a voltage follower (VF) is mandatory for each capacitor C_i.

3.2 Non-Saturated CMOS Cells

These alternative cells use the same Log companding F function as the Saturated SD type given in (4.30), so that both compressor and expander basic building blocks can be taken directly from Figure 4.10.

However, in this case each coefficient of the SD differential equation in (4.28) is split as shown for the A_{ij} case:

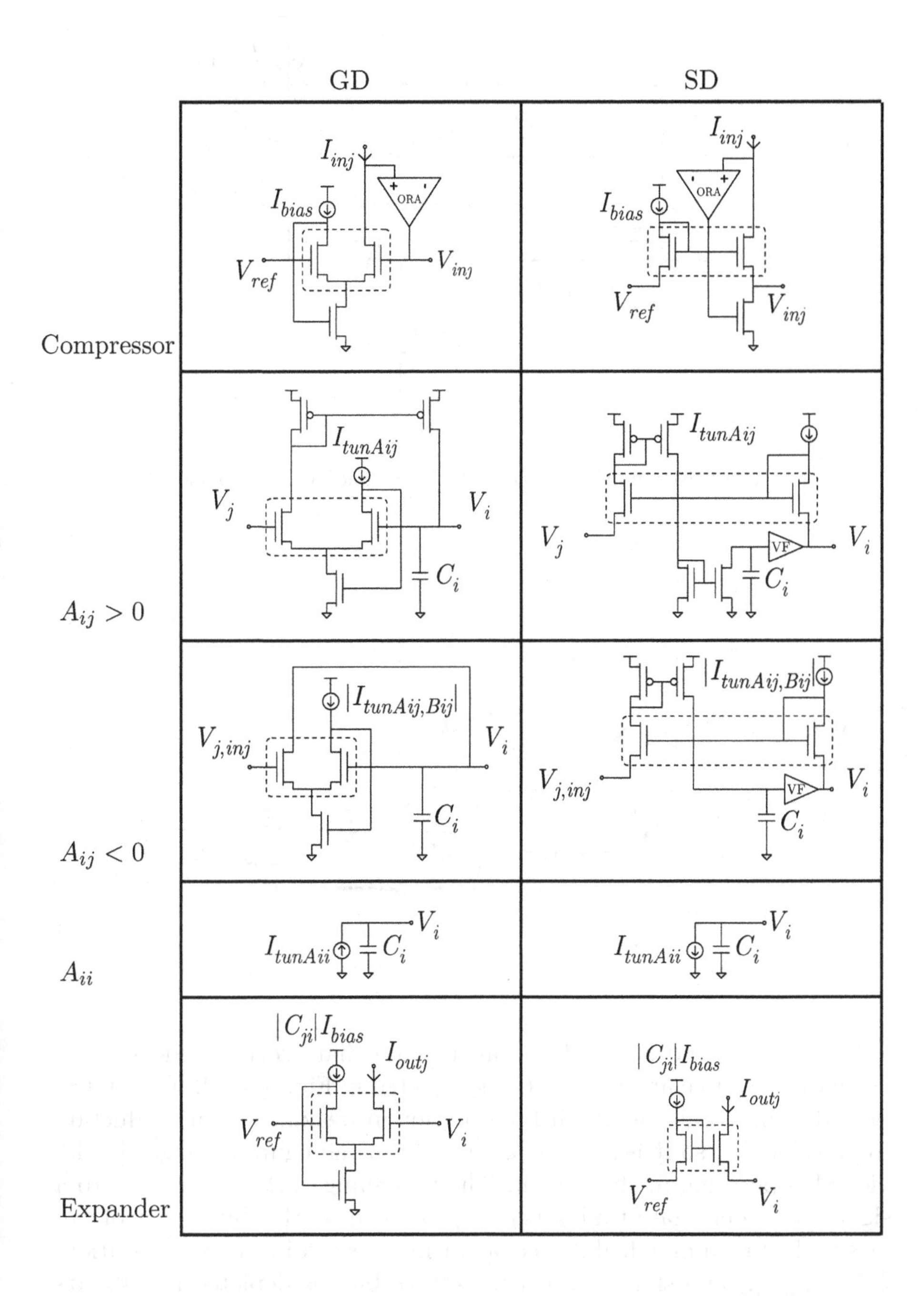

Figure 4.10. Summary of proposed low-voltage basic building blocks for saturated GD and SD NMOS synthesis. Complementary PMOS structures can be derived just by duality.

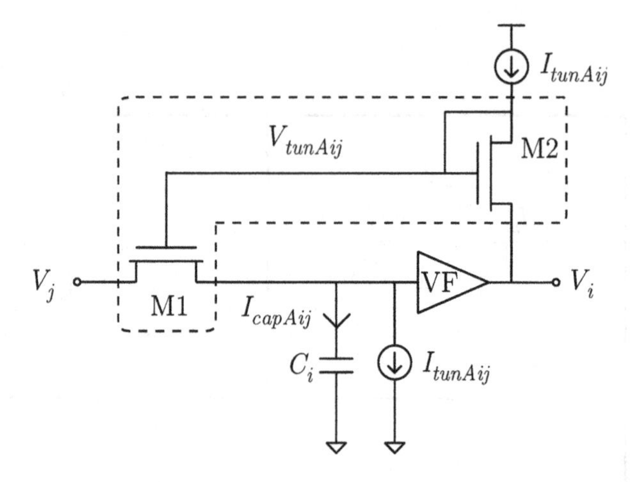

Figure 4.11. New non-saturated NMOS principle for $A_{ij} > 0$ and $i \neq j$.

$$\frac{\mathrm{d}Q_i}{\mathrm{d}t} = C_i\frac{\mathrm{d}V_i}{\mathrm{d}t} = -\sum_{j=1}^{N} I_{tunAij}e^{\frac{V_i-V_j}{U_t}} + \ldots$$

$$\equiv \sum_{\substack{j=1\\j\neq i}}^{N} I_{tunAij}e^{\frac{V_i}{U_t}}\underbrace{\left(e^{\frac{-V_i}{U_t}} - e^{\frac{-V_j}{U_t}}\right)} - \sum_{j=1}^{N} I_{tunAij} + \ldots \tag{4.31}$$

Now, the second summation has an immediate correspondence to a DC current source attached to C_i as depicted in Figure 4.12. The underbraced terms recall the MOSFET equation in weak inversion conduction from Table 2.1, so it is implemented in the same figure through the device M1 operating in this region. The remaining part is equivalent to a signal-dependent gate tuning (V_{tunAij}) applied to M1. Such control can be supplied by a matched device operating in weak inversion saturation, fed at I_{tunAij} and sharing the same source bias as depicted by M2. Its correct functionality requires both isothermal and technological matching for the boxed pair, so this cell is still compatible with Figure 4.2.

The main design equations for M1 and M2 are:

$$\left.\begin{array}{l} I_{D1} = I_S e^{\frac{V_{tunAij}-V_{TO}}{nU_t}} \left(e^{\frac{-V_i}{U_t}} - e^{\frac{-V_j}{U_t}} \right) \\ \\ I_{D2} = I_S e^{\frac{V_{tunAij}-V_{TO}}{nU_t}} e^{\frac{-V_i}{U_t}} \end{array} \right\} I_{D1} \equiv e^{\frac{V_i}{U_t}} \left(e^{\frac{-V_i}{U_t}} - e^{\frac{-V_j}{U_t}} \right) I_{D2} \tag{4.32}$$

Knowing the boundary condition:

$$I_{D2} = I_{tunAij} \tag{4.33}$$

the KCL at the capacitor terminal is identical to the original one in (4.31):

$$I_{capAij} = I_{D1Aij} - I_{tunAij} \equiv -I_{tunAij} e^{\frac{V_i - V_j}{U_t}} \tag{4.34}$$

Based on this idea, an alternative set of low-voltage CMOS cells are proposed in Figure 4.12. As expected, the tuning parameter I_{tunAij} matches with the corresponding saturated SD equation in (4.29). Again, a VF block is required to sense each grounded capacitor C_i, and also a current inverter (CI) may be needed in this case for negative coefficients.

At first glance, this strategy seems to suffer from limited dynamic range due to the need of keeping the main transistor out of saturation. But such condition has not been supposed in the splitting of expression (4.31). In fact, the basic building blocks of Figure 4.12 work perfectly even after enetring the saturation region, which can be understood just as an asymptotic case of the general conduction expression of Table 2.1. This idea can be clearly seen in the graphical comparison of Figure 4.14. In our particular case, saturation is equivalent to blocking the corresponding SD cell of Figure 4.10. In practice, this limit is difficult to reach due to the inner voltage compression of Log companding. For example, taking the SD F function from (4.30), the resulting compressed swing for a 90% full-scale input signal is reduced to:

$$\Delta V_{inj} = U_t \ln \frac{1.9 I_{bias}}{0.1 I_{bias}} \simeq 3U_t \tag{4.35}$$

To illustrate this, different modulation indexes are represented in Figure 4.13, where non-saturation is the usual region of operation for signals of up to 100% full-scale.

In any case, the design of the filter operating point is of essential importance for these cells to be able to take advantage of such available dynamic range and to eliminate redundant parts, as discussed in Section 4.

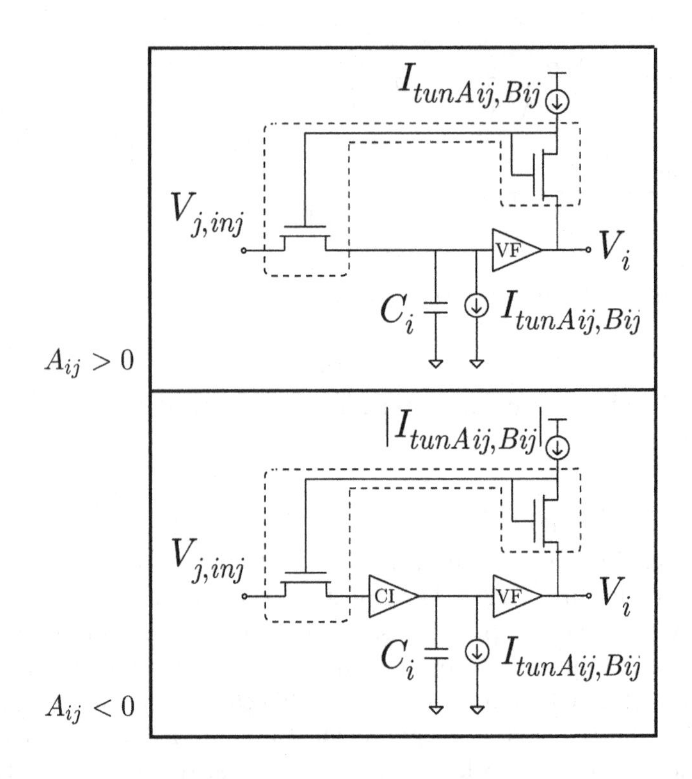

Figure 4.12. Proposed low-voltage non-saturated NMOS cells. Complementary PMOS structures can be derived just by duality.

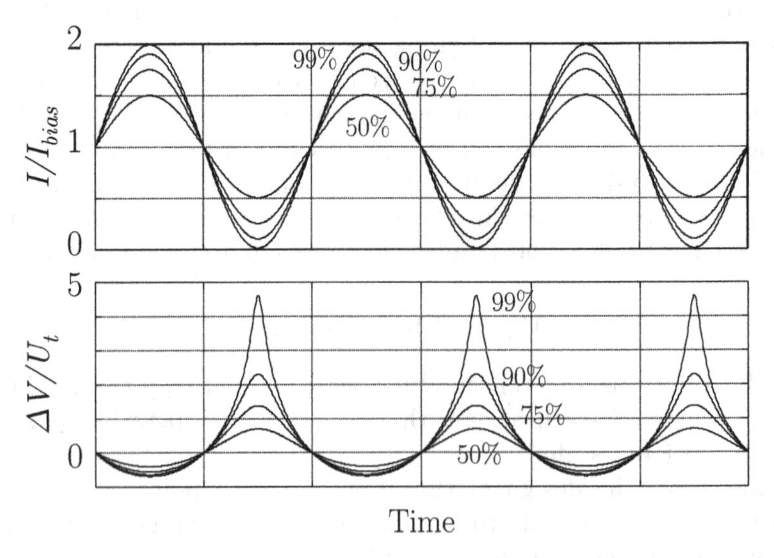

Figure 4.13. SD compression signal boundaries for different modulation indexes.

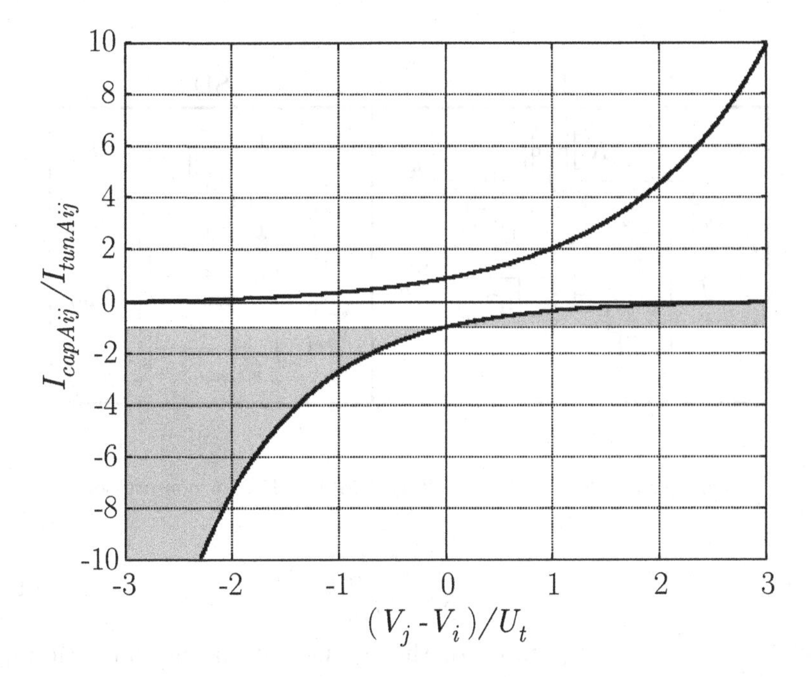

Figure 4.14. Non-linear internal transconductances for GD with $n = 1.3$ (upper) and SD (lower) companding laws. Filled regions indicate contributions from devices in conduction for non-saturated SD cells.

3.3 Auxiliary Circuitry

Usually, compressor and expander MOS pairs must be designed with wide aspect ratios[1], typically $(W/L) > 100$, for practical full-scale values of $I_{max} \in (1\mu\text{A}, 10\mu\text{A})$. Such geometries are not reached in the transconductance network due to the target spectrum $f_{max} < 100\text{KHz}$ and the integrated capacitance values $C_i \in (10\text{pF}, 100\text{pF})$. As a result, a minimum channel length is preferable for compressors and expanders, if allowed by flicker noise and technology mismatching, in order to obtain compact circuits. Similar to the input impedance control presented for amplifiers in Chapter 3, the proposed ORA implementations of Figure 4.15 can be used for compressors to neglect distortion caused by channel length modulation (CLM). Apart from its low-voltage capabilities, such topologies allow for simple frequency compensation according to the damping factor expression:

[1]In order to obtain large I_S values and keep devices in their weak inversion region as studied in Chapter 7.

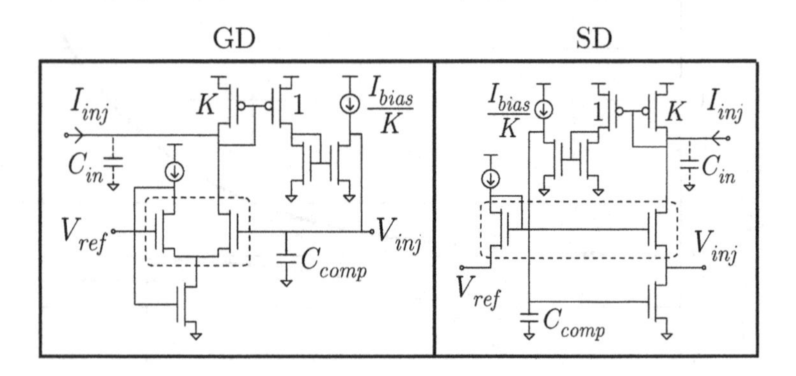

Figure 4.15. Proposed low-voltage CMOS ORA for compressors.

$$\zeta = \frac{1}{2}\sqrt{\frac{KC_{comp}}{C_{in}}} \tag{4.36}$$

where C_{comp} and C_{in} stand for the compensation and parasitic input capacitances, respectively. distortion at the expander is also minimized provided that similar structures exist at the input of the cascaded stage.

Based again on the internal voltage compression, both VF and CI auxiliary controls do not require rail-to-rail operation. In this sense, two compact VF and CI+VF implementations are presented in Figure 4.16 (left) and (right), respectively. In the second auxiliary circuitry, the role of the current inverter (CI) is to supply a low-impedance voltage copy from the Y→X ports and a current copy from the X→Y ports, as symbolized by the following matrix description:

$$\begin{bmatrix} V_X \\ I_Y \end{bmatrix} = \begin{bmatrix} 1 & 0 \\ 0 & 1 \end{bmatrix} \begin{bmatrix} V_Y \\ I_X \end{bmatrix} \tag{4.37}$$

Finally, some additional circuitry is needed to supply V_{ref} levels, which can be understood as operating point shifters in the compressed domain. The built-in generator proposed in Figure 4.17 allows optimization of low-voltage operation by setting the minimum drain drop across the telescopic transistors of Figures 4.10 and 4.12 to:

$$V_{ref(SD)} = U_t \ln(1 + P) \tag{4.38}$$

where upper and lower MOS devices are supposed to operate in weak inversion saturation and conduction, respectively.

A brief comparison between all three types of CMOS basic building blocks proposed in this section results in:

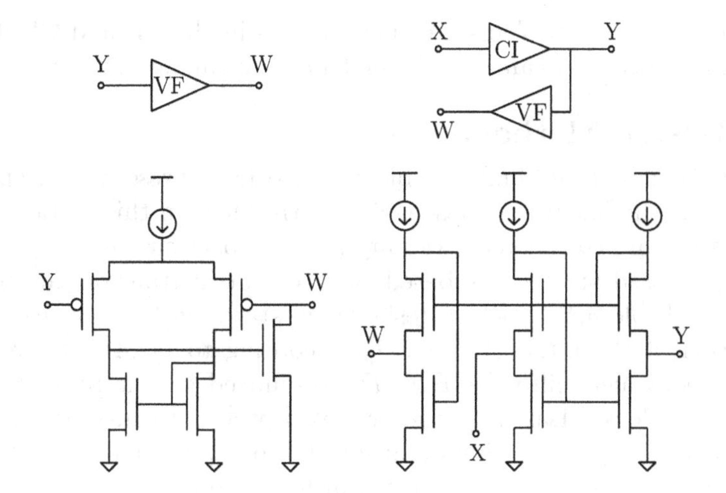

Figure 4.16. Proposed low-voltage VF (left) and VF+CI (right) blocks.

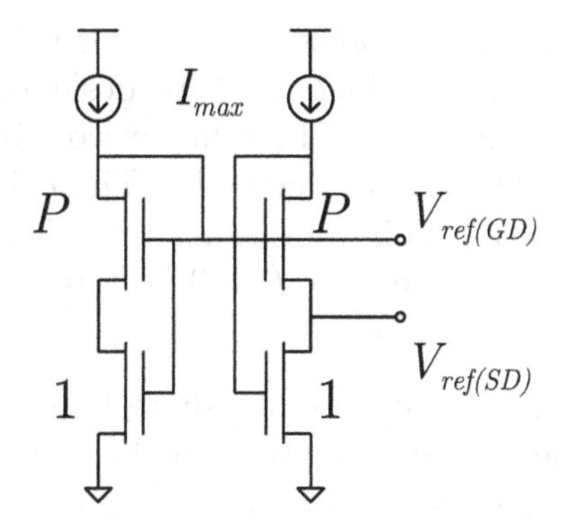

Figure 4.17. Proposed low-voltage CMOS V_{ref} generator.

GD cells: Filter structures synthesized with these elements exhibit the best compacted Si area since no auxiliary circuitry (i.e. VF and CI blocks) is needed. On the other hand, global matching between all cell pairs requires more complex layout techniques.

Saturated SD cells: The main advantage of these basic building blocks is their tuning independence from technology, as well as an easier layout with local matching only at cell level. On the other hand, filter implementations require some extra area and power overhead proportional to their order due to the auxiliary circuitry.

Non-saturated GD cells: The same considerations from the previous class also applies for these blocks. Furthermore, DC errors are reduced in this case due to its direct drain-to-source connectivity.

Since matching problems are then located in the VF and CI blocks, their optimization can be independent from tuning.

4. Design Methodology

Once the new basic building blocks have been presented, a specific synthesis methodology for these cells is introduced in this section. Two design issues are considered: stability and compact circuits.

The proposed strategy is based on the matrix transformations described in Table 4.1. Such a design flow starts with an arbitrary SS description of the filter $[A, B, C, D]_0$ according to (4.24), and returns a totally equivalent filter $[A, B, C, D]_2$ optimized for the presented basic building blocks. Basically, the proposed procedure ensures the same compressed voltage operating point for all inputs, internal variables and outputs through a two-step matrix transformation.

The main advantages of synthesizing filters under Table 4.1 are:

- A stable and centered compressed operating point. Due to the Log companding function F defined in (1.5), designers must ensure that $Y \in \mathbb{R}^+$, otherwise $X \to \infty$. In other words, if $I \to 0$ then $V \to \mp\infty$ for GD and SD laws, respectively. The existence of a positive and centered operating point solution in the I-domain through this procedure does not ensure that $I > 0$ is fulfilled $\forall t$, but it helps to find a stable operating point.

- All compressors and expanders can share half of the circuitry. Following step 1, the circuit reductions of Figure 4.18 can be applied between input compressors and output expanders. The resulting Si area savings are of special interest due to the geometry requirements of these blocks as explained in Chapter 7.

- Simplification of internal transconductances. Step 2 (or alternatively 3) in Table 4.1 ensures the same operating point for all inputs, internal SS variables and outputs. This situation usually returns duplicated or at least non-independent tuning coefficients in matrix A and B. In that case, circuit reductions based on simple geometrical MOS ratios can be applied, like the proposals of Figure 4.19 where:

$$I_{DCi} = \frac{I_{tunAik}}{|A_{ik}|} \left[\sum_{j=1}^{N} A_{ij} + \sum_{j=1}^{M} B_{ij} \right] \tag{4.39}$$

- Suppression of all current sources attached to capacitors in Figure 4.12. Equal operating points from Table 4.1 automatically remove all DC

Table 4.1. Proposed SS matrix algorithm for filter synthesis.

---

0 Initial matrix description A_0, B_0, C_0 and $D_0 \equiv 0$.

1 Normalize C through a linear transformation (M_{norm}) to share circuitry and DR_I for compressors/expanders, and $Y_{out}(DC) \equiv Y(DC)$:

$$C_0 M_{norm}^{-1} \doteq I \quad \xrightarrow[M_{norm}]{} \quad A_1, B_1, C_1 \equiv I \text{ and } D_1 \equiv 0$$

2 Search for a linear transformation (M_{op}) to keep previous C_1 and achieve $Y(DC) \equiv Y_{in}(DC)$:

$$C_2 = C_1 M_{op}^{-1} \doteq C_1 \qquad\qquad C_1 = M_{op} C_1$$

$$\bar{0}_{order} = A_2 \bar{1}_{order} y_{bias} + B_2 \bar{1}_{inputs} y_{bias} \qquad \bar{1}_{order} = -M_{op} A_1^{-1} B_1 \bar{1}_{inputs}$$

$$\Big\downarrow M_{op}$$

$$A_2, B_2, C_2 \equiv C_1 \text{ and } D_2 \equiv 0$$

E.g. One-input second-order systems:

$$\begin{bmatrix} 1 & 0 \end{bmatrix} = M_{op} \begin{bmatrix} 1 & 0 \end{bmatrix}$$

$$M_{op} = \begin{bmatrix} 1 & 0 \\ m_{21} & m_{22} \end{bmatrix} \quad \begin{bmatrix} 1 \\ 1 \end{bmatrix} = - \begin{bmatrix} 1 & 0 \\ m_{21} & m_{22} \end{bmatrix} A_1^{-1} B_1$$

3 If $\nexists M_{op}$, then use an extra dummy input to verify $Y(DC) \equiv Y_{in}(DC)$:

$$\bar{0}_{order} = A_1 \bar{1}_{order} + B_1 \bar{1}_{inputs} + \bar{B}_{dummy} \qquad B_2 = \left[B_1 | \bar{B}_{dummy} \right]$$

$$\Big\downarrow$$

$$A_1, B_2, C_1 \text{ and } D_1 \equiv 0$$

E.g. One-input second-order systems:

$$\bar{B}_{dummy} = -A_1 \begin{bmatrix} 1 \\ 1 \end{bmatrix} - B_1$$

Note: if more than one solution is possible in steps 2 or 3, then choose to obtain same coefficients per row or positive coefficients outside diagonal which simplify non-linear transconductance cells.

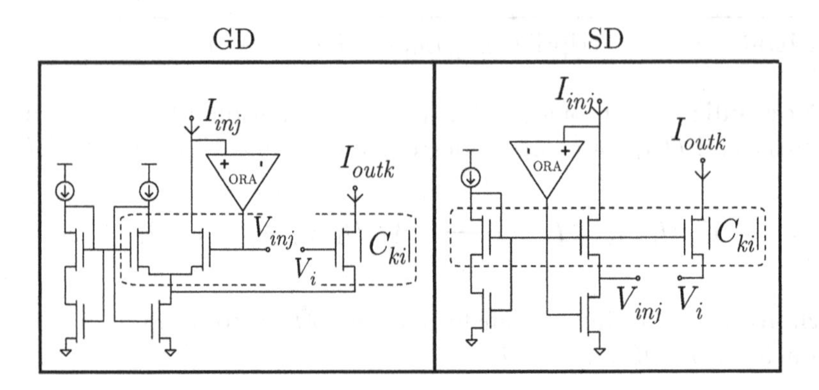

Figure 4.18. Compressor and expander simplification.

current sources for the non-saturated SD cells. This fact can be easily verified just by forcing the operating point condition $\mathrm{d}I_i/\mathrm{d}t \doteq 0$ in (4.25):

$$0 = \sum_{j=1}^{N} A_{ij} I_j(DC) + \sum_{j=1}^{M} B_{ij} I_{inj}(DC) \qquad \text{for } i\text{=1 to } N \tag{4.40}$$

Since all $I(DC)$ are identical, (4.39) results in $I_{DCi} \equiv 0$. Qualitatively, such a simplification comes from the fact that since all internal compressed voltages exhibit the same DC value, no static current can flow through the network of non-saturated MOS devices. Apart from the circuit reduction itself, the above characteristic optimizes the filter full-scale.

5. Case Studies

The following subsections include the synthesis of some basic filter prototypes. They show both the usage of the three proposed types of CMOS basic building blocks as well as the application of Table 4.1.

5.1 Integrator

The general transfer function and its equivalent matrix description can be written as:

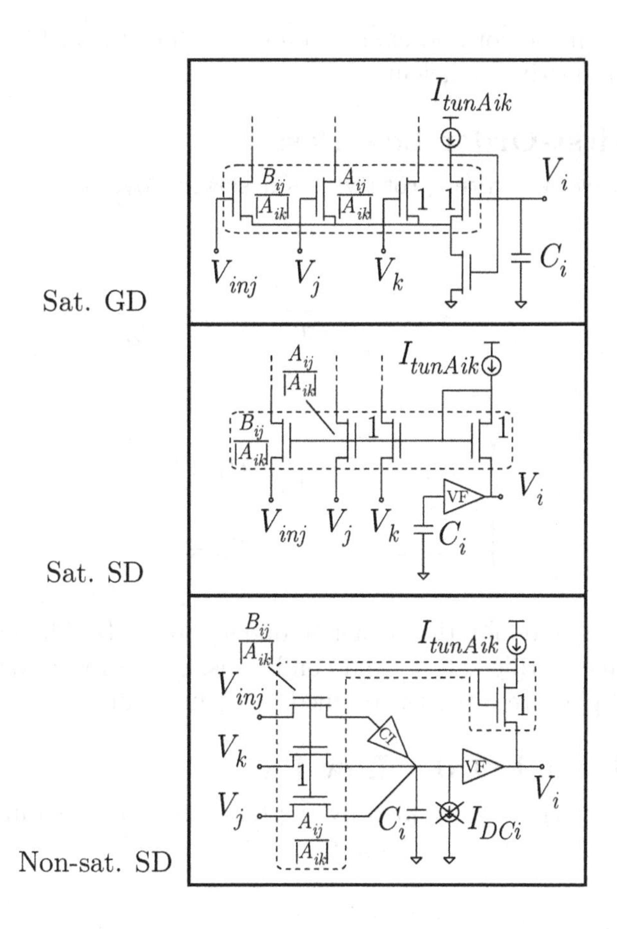

Figure 4.19. Internal coefficient reductions.

$$H(s) = \frac{w_o}{s} \qquad \tau = \frac{1}{w_o} \tag{4.41}$$

$$\left\{ \begin{array}{ll} A_0 = \left[\, 0 \,\right] & B_0 = \left[\, w_o \,\right] \\ \\ C_0 = \left[\, 1 \,\right] & D_0 = \left[\, 0 \,\right] \end{array} \right. \tag{4.42}$$

where $w_o/2\pi$ and τ stand for the Natural frequency and time constant, respectively. The above SS model already verifies Table 4.1 and can be directly used to synthesize the topologies depicted in Figure 4.20.

Note that compressor and expander share half of the cell because of the unitary C_0 matrix coefficient.

5.2 First-Order Low-Pass

In this case, the filter prototype is described by:

$$H(s) = \frac{w_o}{s + w_o} \qquad f_c = \frac{w_o}{2\pi} \tag{4.43}$$

$$\begin{cases} A_0 = \begin{bmatrix} -w_o \end{bmatrix} & B_0 = \begin{bmatrix} w_o \end{bmatrix} \\ C_0 = \begin{bmatrix} 1 \end{bmatrix} & D_0 = \begin{bmatrix} 0 \end{bmatrix} \end{cases} \tag{4.44}$$

where f_c stands for the corner frequency at -3dB. The above SS description also verifies all desired conditions, so that the first-order low-pass topologies are directly obtained in Figure 4.21.

5.3 Second-Order Low-Pass

This filter type is described by the following transfer function prototype:

$$H(s) = \frac{w_o^2}{s^2 + 2\zeta w_o s + w_o^2} \qquad f_c = \frac{w_o}{2\pi}\sqrt{\sqrt{(2\zeta^2 - 1)^2 + 1} - (2\zeta^2 - 1)}$$

$$\text{for } \zeta < \frac{1}{\sqrt{2}} \qquad \begin{cases} f_{peak} = \dfrac{w_o}{2\pi}\sqrt{1 - 2\zeta^2} \\ |H(j2\pi f_{peak})| = \dfrac{1}{2\zeta\sqrt{1 - \zeta^2}} \end{cases} \tag{4.45}$$

where $w_o/2\pi$, ζ, f_c and f_{peak} stand for the central frequency, damping factor, corner frequency and peak frequency, respectively. A valid SS description can be derived through standard techniques [14]:

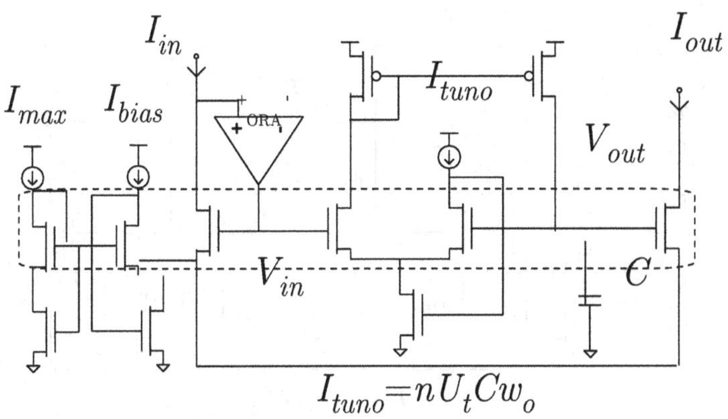

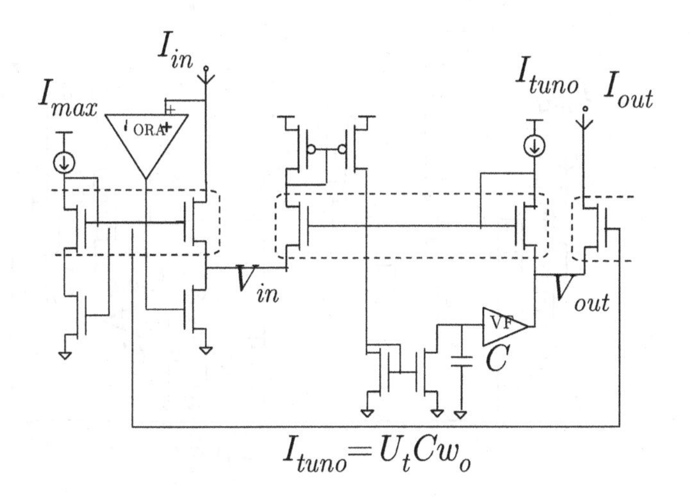

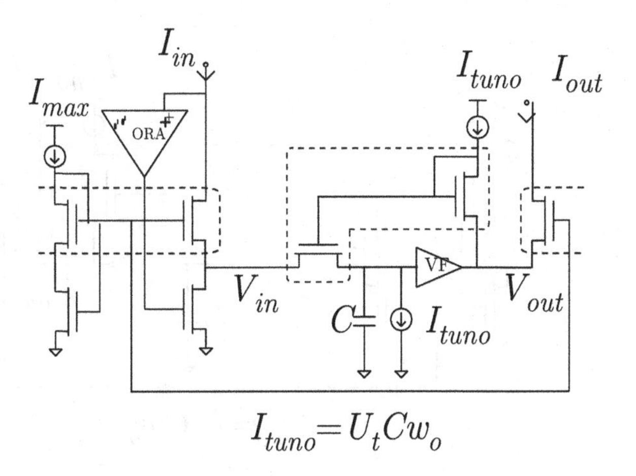

Figure 4.20. The integrator stage implemented using GD (upper), Saturated SD (middle) and non-saturated SD (lower) CMOS cells.

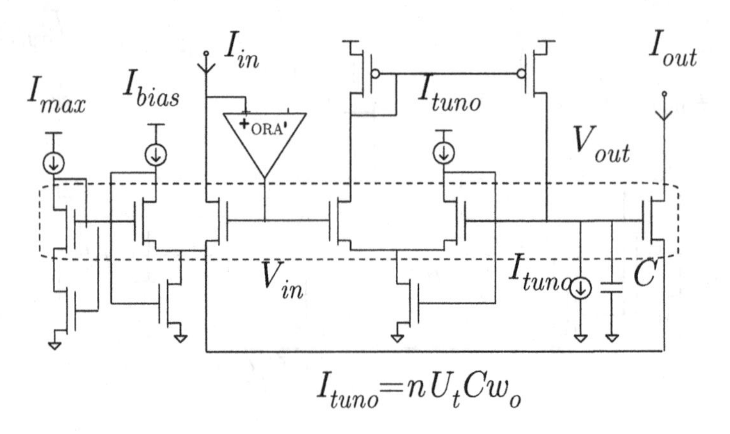

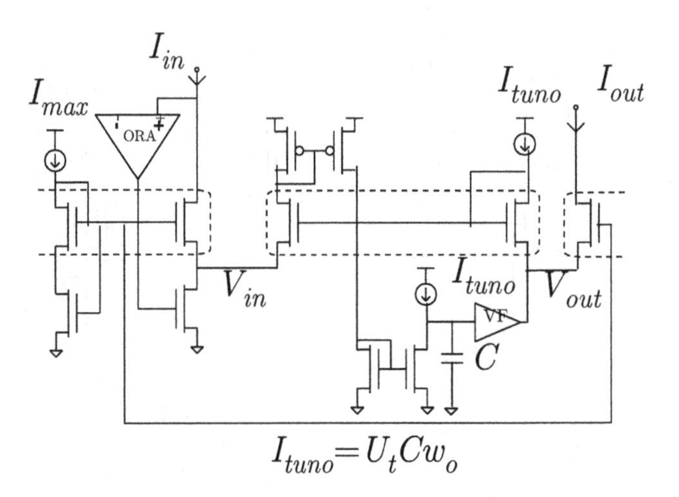

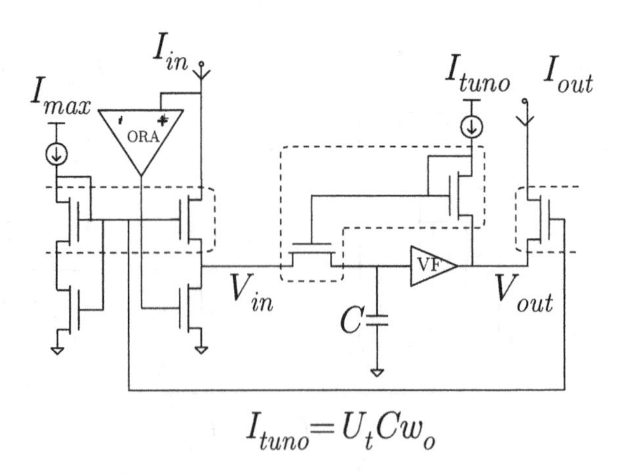

Figure 4.21. First-order low-pass saturated GD (upper), saturated SD (middle) and non-saturated SD (lower) implementations.

$$\begin{cases} A_0 = \begin{bmatrix} -2\zeta w_o & -w_o \\ w_o & 0 \end{bmatrix} & B_0 = \begin{bmatrix} 0 \\ -w_o \end{bmatrix} \\ C_0 = \begin{bmatrix} 1 & 0 \end{bmatrix} & D_0 = \begin{bmatrix} 0 \end{bmatrix} \end{cases} \tag{4.46}$$

Clearly, this initial matrix system already verifies step 1 of Table 4.1. However, it fails to ensure a positive stable operating point for the first SS variable since all coefficients are negative and all variables must be positive. In this sense, some matrix transformation is required to ensure a DC solution in the Log domain. After applying Table 4.1 to (4.46) as detailed in Table 4.2, the resulting SS system $[A, B, C, D]_2$ still represents the same filter prototype (4.45) but now it allows a positive and centered operating point for all input, internal and output variables. Since the A and B matrices present equal coefficients per row, the nonlinear transconductance network can be strongly simplified as depicted in Figures 4.22, 4.23 and 4.24. Furthermore, independent tuning of the Central frequency and dumping factor may be easily implemented through the control of the corresponding tuning currents.

5.4 Second-Order Band-Pass

This filter prototype is initially described by the following transfer function and SS matrix system:

$$H(s) = \frac{\frac{w_o}{Q} s}{s^2 + \frac{w_o}{Q} s + w_o^2} \qquad f_{\substack{upper \\ lower}}(-3\text{dB}) = \frac{1}{2\pi}\frac{w_o}{2Q}\left(\sqrt{4Q^2+1} \pm 1\right)$$

$$BW(-3\text{dB}) = \frac{1}{2\pi}\frac{w_o}{Q}$$

$$f_{upper} f_{lower} = \left(\frac{w_0}{2\pi}\right)^2 \tag{4.47}$$

$$\begin{cases} A_0 = \begin{bmatrix} -\frac{w_o}{Q} & -w_o \\ w_o & 0 \end{bmatrix} & B_0 = \begin{bmatrix} w_o \\ 0 \end{bmatrix} \\ C_0 = \begin{bmatrix} \frac{1}{Q} & 0 \end{bmatrix} & D_0 = \begin{bmatrix} 0 \end{bmatrix} \end{cases} \tag{4.48}$$

Table 4.2. One-input one-output 2nd order low-pass example of Table 4.1.

---

0 Initial matrix description (4.46).

1 Normalization of C_0 is already verified. Hence:

$$A_1 \equiv A_0 \qquad B_1 \equiv B_0 \qquad C_1 \equiv C_0 \qquad D_1 \equiv D_0$$

2 Search for a linear transformation (M_{op}) to keep previous C_1 and achieve $Y(DC) \equiv Y_{in}(DC)$:

$$\begin{bmatrix} 1 \\ 1 \end{bmatrix} = -M_{op}A_1^{-1}B_1 = -\begin{bmatrix} 1 & 0 \\ m_{21} & m_{22} \end{bmatrix} \begin{bmatrix} 0 & 1/w_o \\ -1/w_o & -2\zeta/w_o \end{bmatrix} \begin{bmatrix} 0 \\ -w_o \end{bmatrix}$$

$$m_{21} - 2\zeta m_{22} = 1$$

3 Choose the most suitable M_{op} in terms of circuit implementation:

$$A_2 = \begin{bmatrix} \left(-2\zeta + \frac{m_{21}}{m_{22}}\right) w_o & -\frac{w_o}{m_{22}} \\ \left(-2\zeta m_{21} + \frac{m_{21}^2}{m_{22}} + m_{22}\right) w_o & -\frac{m_{21}}{m_{22}} w_o \end{bmatrix}$$

$$B_2 = \begin{bmatrix} 0 \\ -m_{22}w_o \end{bmatrix}$$

The simplest solution by inspection is $m_{21} \equiv 0 \Longrightarrow M_{op} = \begin{bmatrix} 1 & 0 \\ 0 & -1/2\zeta \end{bmatrix}$,

thus:

$$A_2 = \begin{bmatrix} -2\zeta w_o & 2\zeta w_o \\ -\frac{w_o}{2\zeta} & 0 \end{bmatrix} \qquad B_2 = \begin{bmatrix} 0 \\ \frac{w_o}{2\zeta} \end{bmatrix}$$

$$C_2 = \begin{bmatrix} 1 & 0 \end{bmatrix} \qquad D_2 = \begin{bmatrix} 0 \end{bmatrix}$$

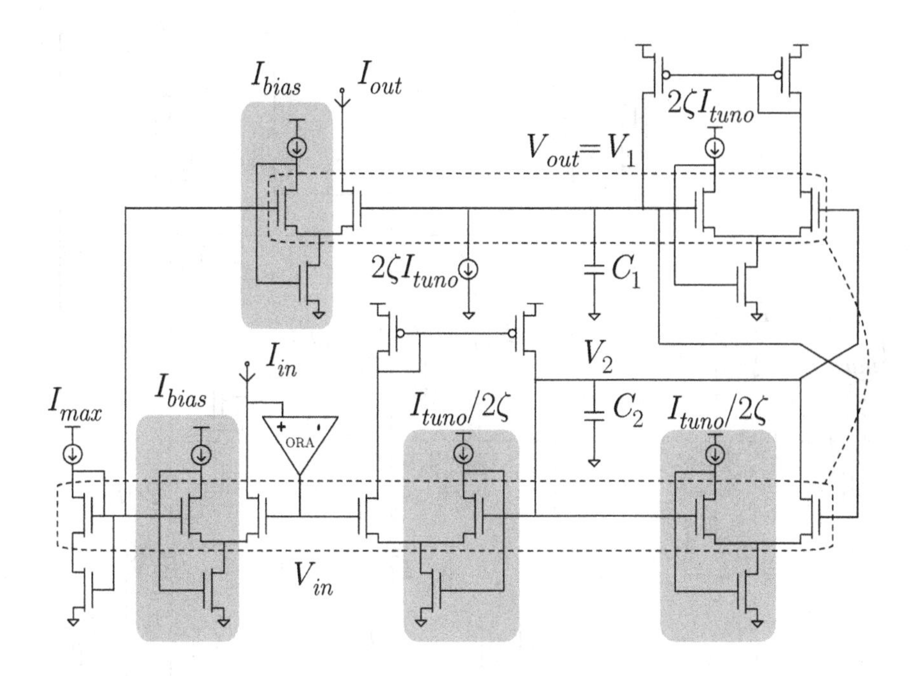

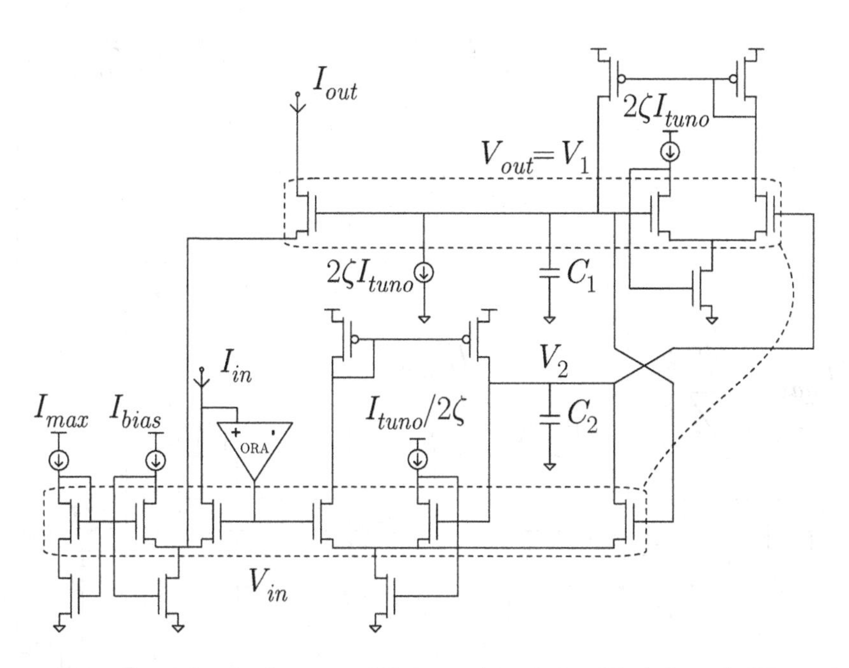

Figure 4.22. Second-order low-pass full (upper) and simplified (lower) GD implementations with $C_1 = C_2 \doteq C$ and $I_{tuno} = nU_tCw_o$.

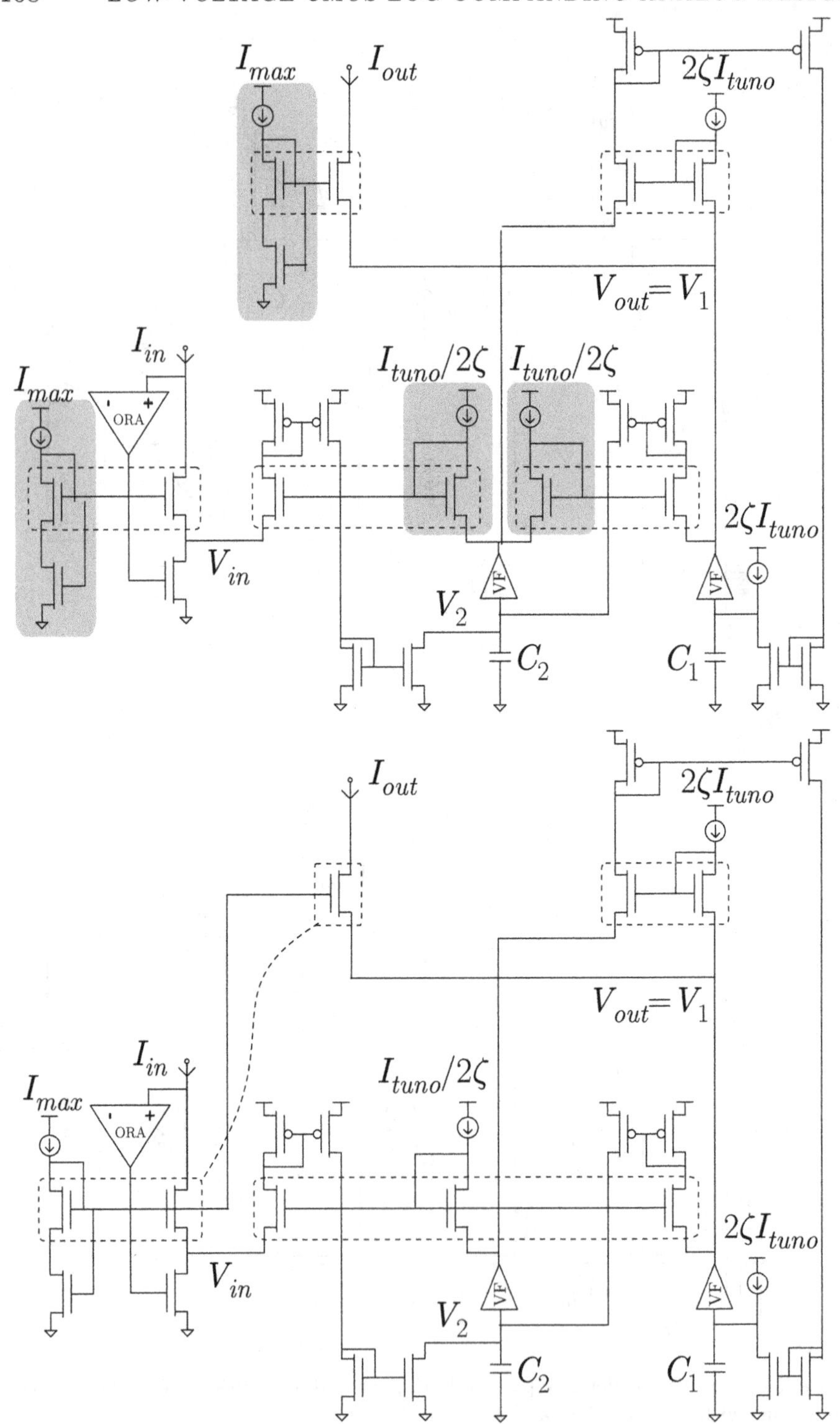

Figure 4.23. Second-order low-pass full (upper) and simplified (lower) saturated SD implementations with $C_1 = C_2 \doteq C$ and $I_{tuno} = U_t C w_o$.

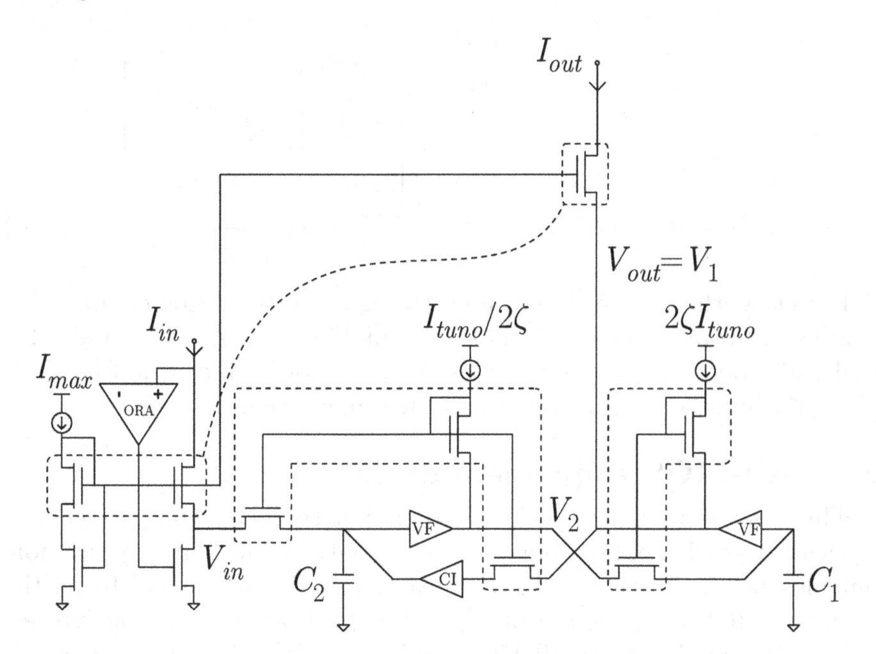

Figure 4.24. Second-order low-pass simplified non-saturated SD implementation with $C_1 = C_2 \doteq C$ and $I_{tuno} = U_t C w_o$.

where $w_o/2\pi$, Q, f_{lower}, f_{upper} and BW stand for the central frequency, quality factor, lower and upper frequencies, and bandwidth, respectively. Again, a stability problem arises, related in this case to the second SS variable. Hence, Table 4.1 is applied to $[A, B, C, D]_0$ in Table 4.3. However, unlike the previous filter prototype, no linear transformation exists to satisfy all requirements, so the extra dummy alternative (i.e. step 3) is chosen in the same procedure. Again, the resulting SS system $[A, B, C, D]_2$ exhibits a stable and centered operating point, shares parts between compressor and expander, and has equal coefficients per row which simplifies the non-linear transconductances. Furthermore, the second-order low-pass prototype of (4.45) can be also obtained by adding an expander to the second SS internal variable:

$$\left\{ \begin{array}{ll} A_3 = \begin{bmatrix} -\frac{w_o}{Q} & -\frac{w_o}{Q} \\ Qw_o & 0 \end{bmatrix} & B_3 = \begin{bmatrix} \frac{w_o}{Q} & \frac{w_o}{Q} \\ 0 & -Qw_o \end{bmatrix} \\ & \\ C_3 = \begin{bmatrix} 1 & 0 \\ 0 & 1 \end{bmatrix} & D_3 = \begin{bmatrix} 0 & 0 \\ 0 & 0 \end{bmatrix} \end{array} \right. \tag{4.49}$$

Now, both second-order band- and low-pass transfer functions are obtained for the same input as marked in:

$$H_3(s) = C_3(sI - A_3)^{-1}B_3 + D_3 = \frac{\begin{bmatrix} \frac{w_o}{Q}s & \frac{w_o}{Q}s + w_o^2 \\ w_o^2 & -Qw_os \end{bmatrix}}{s^2 + \frac{w_o}{Q}s + w_o^2} \tag{4.50}$$

Direct synthesis of such a system through the three Log companding CMOS basic building blocks is shown in Figures 4.25, 4.26 and 4.27. Full and independent tuning has been chosen for both central frequency and quality factor parameters in all implementations.

6. All-MOS Implementations

The basic idea of this section is to exploit the inner voltage dynamic range compression of Log companding not only for low-voltage operation, but also to allow the use of non-linear capacitive elements. In particular, the MOS implementation of such capacitors is of special interest in order to finally obtain all-MOS analog circuit techniques compatible with digital CMOS technologies.

Recalling the original differential equation in the linear domain (4.25) and the two MOSFET companding functions F defined in (4.26), the actual non-linear equation to be implemented in the V-domain is:

$$\frac{\mathrm{d}V_i}{\mathrm{d}t} = \begin{cases} \sum_{j=1}^{N} nU_tA_{ij}e^{\frac{V_j - V_i}{nU_t}} + \sum_{j=1}^{M} nU_tB_{ij}e^{\frac{V_{inj} - V_i}{nU_t}} & \text{GD} \\ -\sum_{j=1}^{N} U_tA_{ij}e^{\frac{V_i - V_j}{U_t}} - \sum_{j=1}^{M} U_tB_{ij}e^{\frac{V_i - V_{inj}}{U_t}} & \text{SD} \end{cases} \tag{4.51}$$

As argued in Section 3, the above expression is usually multiplied by C_i at each side, so that the left-hand side is directly identified in (4.28) as the charge and discharge current (I_{capi}) of the grounded capacitor C_i, which stores the state-space variable V_i. Three complete types of low-voltage CMOS implementations using this approach have already been presented in Section 3.

However, from the point of view of the circuit, the real SS variable, controllable through I_{capi}, is the charge stored in the capacitor (Q_i). In general, such a variation can be expressed in terms of V_i as:

$$I_{capi} = \frac{\mathrm{d}Q_i}{\mathrm{d}t} = C_i\frac{\mathrm{d}V_i}{\mathrm{d}t} + \frac{\mathrm{d}C_i}{\mathrm{d}t}V_i = \left(C_i + \frac{\mathrm{d}C_i}{\mathrm{d}V_i}V_i\right)\frac{\mathrm{d}V_i}{\mathrm{d}t} \tag{4.52}$$

Table 4.3. One-input 2nd order band-pass example for Table 4.1.

0 Initial matrix description (4.48).

1 Normalization of C through $M_{norm} = \begin{bmatrix} 1/Q & 0 \\ 0 & 1 \end{bmatrix}$:

$$A_1 = \begin{bmatrix} -\dfrac{w_o}{Q} & -\dfrac{w_o}{Q} \\ Qw_o & 0 \end{bmatrix} \quad B_1 = \begin{bmatrix} \dfrac{w_o}{Q} \\ 0 \end{bmatrix}$$

$$C_1 = \begin{bmatrix} 1 & 0 \end{bmatrix} \quad D_1 = \begin{bmatrix} 0 \end{bmatrix}$$

2 Search for a linear transformation (M_{op}) to keep previous C_1 and achieve $Y(DC) \equiv Y_{in}(DC)$:

$$\begin{bmatrix} 1 \\ 1 \end{bmatrix} = -M_{op}A_1^{-1}B_1 = -\begin{bmatrix} 1 & 0 \\ m_{21} & m_{22} \end{bmatrix}\begin{bmatrix} 0 & 1/Qw_o \\ -Q/w_o & -1/Qw_o \end{bmatrix}\begin{bmatrix} w_o/Q \\ 0 \end{bmatrix} = \begin{bmatrix} 0 \\ m_{22} \end{bmatrix}$$

3 If $\nexists M_{op}$, then use an extra dummy input to verify $Y(DC) \equiv Y_{in}(DC)$:

$$\bar{B}_{dummy} = -A_1 \begin{bmatrix} 1 \\ 1 \end{bmatrix} - B_1 = \begin{bmatrix} \dfrac{w_o}{Q} \\ -Qw_o \end{bmatrix}$$

Thus, the final description with $B_2 = [B_1 | \bar{B}_{dummy}]$:

$$A_2 = \begin{bmatrix} -\dfrac{w_o}{Q} & -\dfrac{w_o}{Q} \\ Qw_o & 0 \end{bmatrix} \quad B_2 = \begin{bmatrix} \dfrac{w_o}{Q} & \dfrac{w_o}{Q} \\ 0 & -Qw_o \end{bmatrix}$$

$$C_2 = \begin{bmatrix} 1 & 0 \end{bmatrix} \quad D_2 = \begin{bmatrix} 0 & 0 \end{bmatrix}$$

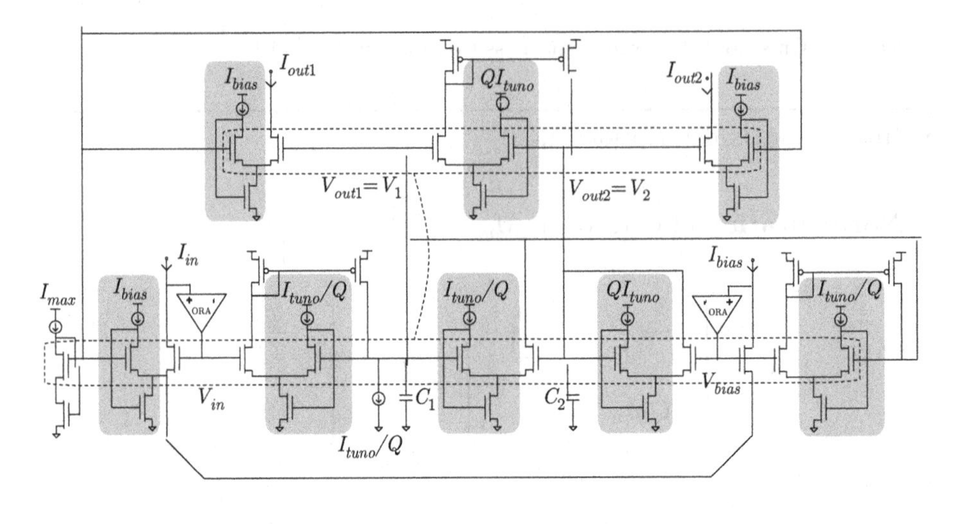

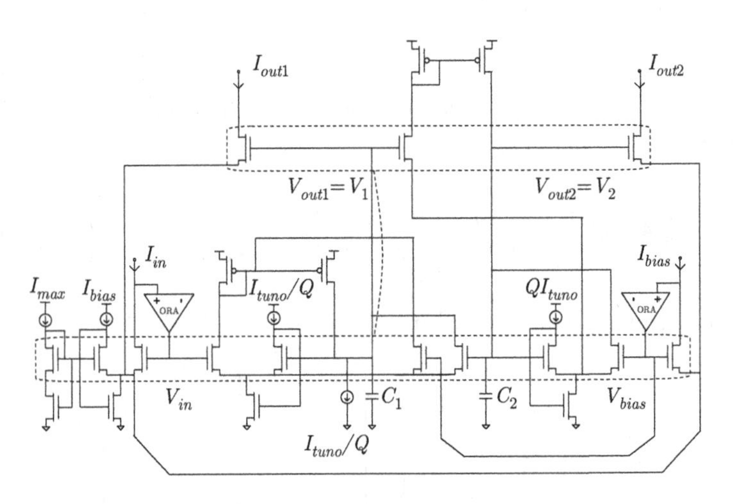

Figure 4.25. Second-order band/low-pass full (upper) and simplified (lower) GD implementations with $C_1 = C_2 \doteq C$ and $I_{tuno} = nU_tCw_o$.

Hence, after comparing the above equation with (4.28), it is easy to conclude that the latter only applies for linear capacitors. In practice, this linearity requirement for the integrated element C_i creates the need for poly-Si capacitors in CMOS technologies.

In case of using non-linear capacitors, two different sources of signal distortion can be derived from (4.52): variation of the capacitance ($\frac{\mathrm{d}C_i}{\mathrm{d}V_i} \neq 0$), and the signal range itself (V_i). In this sense, the instantaneous Log companding processing seems especially suitable due to its internal compression of voltage dynamic range.

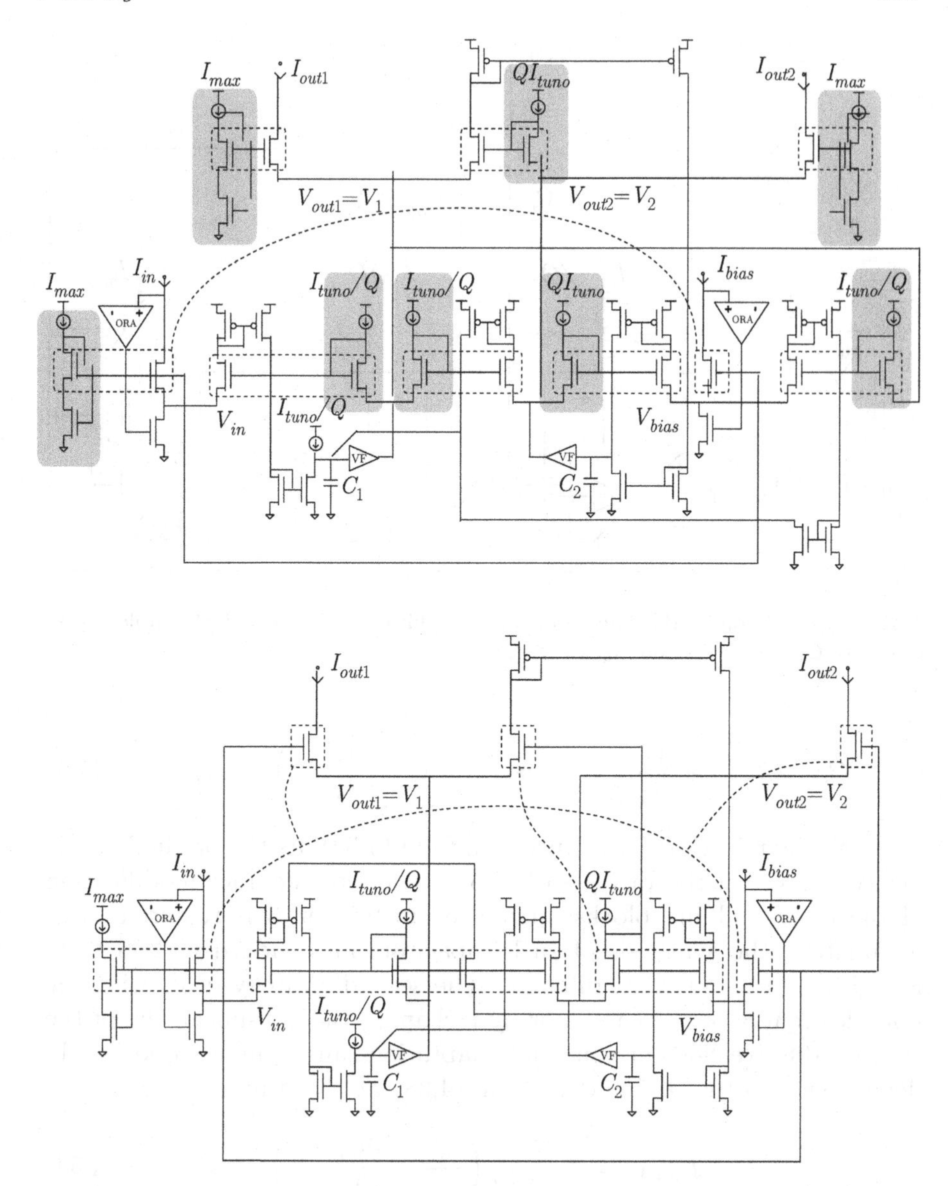

Figure 4.26. Second-order band/low-pass full (upper) and simplified (lower) saturated SD implementations with $C_1 = C_2 \doteq C$ and $I_{tuno} = U_t C w_o$.

Two general solutions may be proposed to make (4.28) and (4.52) compatible: capacitance and transconductance non-linear compensations. The first approach focuses on obtaining a special non-linear capacitance law (e.g. by parallel addition of different elements) verifying:

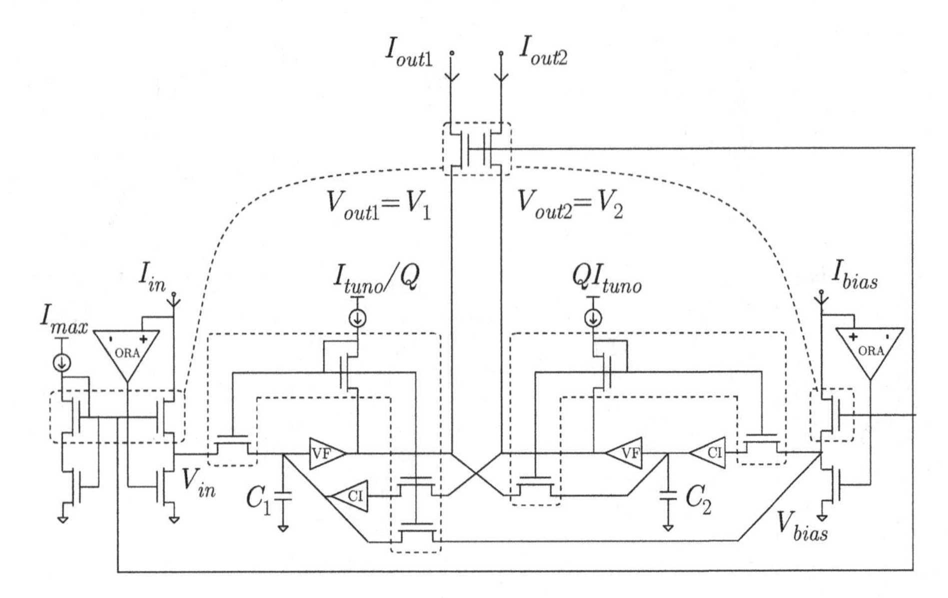

Figure 4.27. Second-order band/low-pass simplified non-saturated SD implementation with $C_1 = C_2 \doteq C$ and $I_{tuno} = U_t C w_o$.

$$C_i = C_{oi}\left(1 + \frac{V_{oi}}{V_i}\right) \quad \Rightarrow \quad \left(C_i + \frac{\mathrm{d}C_i}{\mathrm{d}V_i}V_i\right) \equiv C_{oi} \tag{4.53}$$

Then, the resulting I_{capi} expression from (4.52) fits the original equivalence (4.28) exactly for $C_i \equiv C_{oi}$, so no transconductance modification of the basic building blocks introduced in Section 3 is required. Unfortunately, the synthesis of such capacitors in standard CMOS technologies seems somewhat difficult. The second strategy deals with the non-ideal curve of the capacitor by reshaping the non-linear law of the transconductive element. Such an adaptation can be understood as a V_i dependence on the tuning current in (4.28) resulting in:

$$I_{tunAij} = I_{tunoAij}\left(\frac{C_i}{C_{oi}} + \frac{\mathrm{d}C_i}{\mathrm{d}V_i}\frac{V_i}{C_{oi}}\right) \tag{4.54}$$

where $I_{tunoAij}$ is the original tuning parameter defined by (4.29) for the nominal capacitance at the operating point ($C_i \doteq C_{oi}$). This approach can be locally optimized due to the reduced excursions of V_i in the Log filtering. Notice that practical inner voltage swings for both GD and SD compression laws expressed in (4.30) are typically limited to less than $5U_t$. Furthermore, the transconductance compensation approach is also compatible with the circuit reduction techniques developed in the design methodology of Section 4.

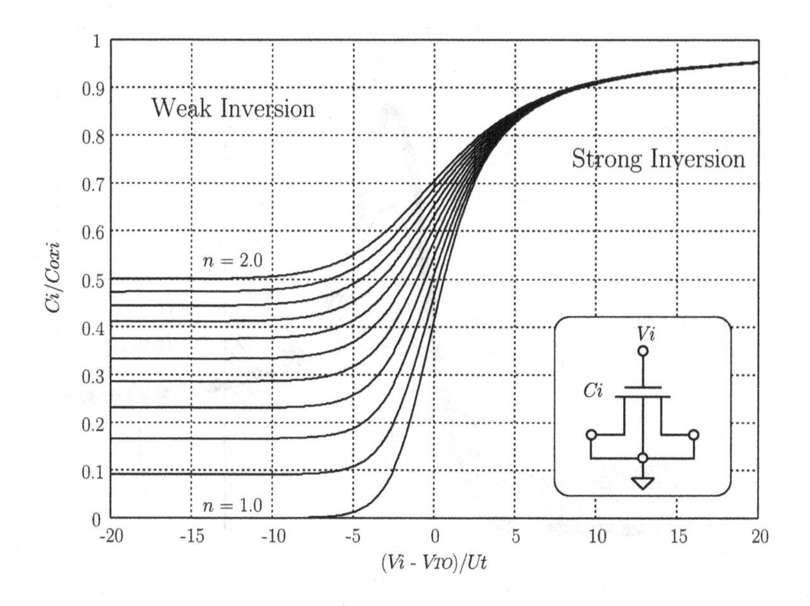

Figure 4.28. MOS gate-to-gate capacitance versus voltage.

Based on this idea, an NMOS non-linear capacitor implementation is proposed according to Figure 4.28 as:

$$C_i \doteq C_{GG} \equiv C_{GB} + C_{GS} + C_{GD} \tag{4.55}$$

According to the capacitive model of Chapter 2 and particularized for $V_{SB} = V_{DB} = 0$, the quasi-static gate-to-gate MOS capacitance for all regions of operation is:

$$C_i = C_{oxi}\frac{\frac{n-1}{n} + 2\sqrt{IC}\left(1 - e^{-\sqrt{IC}}\right)}{1 + 2\sqrt{IC}\left(1 - e^{-\sqrt{IC}}\right)} \simeq \begin{cases} \frac{n-1}{n}C_{oxi} & IC \ll 1 \\ C_{oxi} & IC \gg 1 \end{cases} \tag{4.56}$$

where C_{oxi} stands for the nominal gate oxide capacitance, while the forward (IC_f) and reverse (IC_r) channel inversion coefficients are identical ($IC \doteq IC_f = IC_r$) and related to the stored voltage as follows:

$$IC = \ln^2\left(1 + e^{\frac{V_i - V_{TO}}{2nU_t}}\right) \tag{4.57}$$

The analytical expression of C_i versus V_i is plotted in Figure 4.28. A sweep of $\pm 20U_t$ around the threshold voltage has been chosen, which is equivalent to $V_{TO} = 0.5$V, for 1V supply voltage at room temperature.

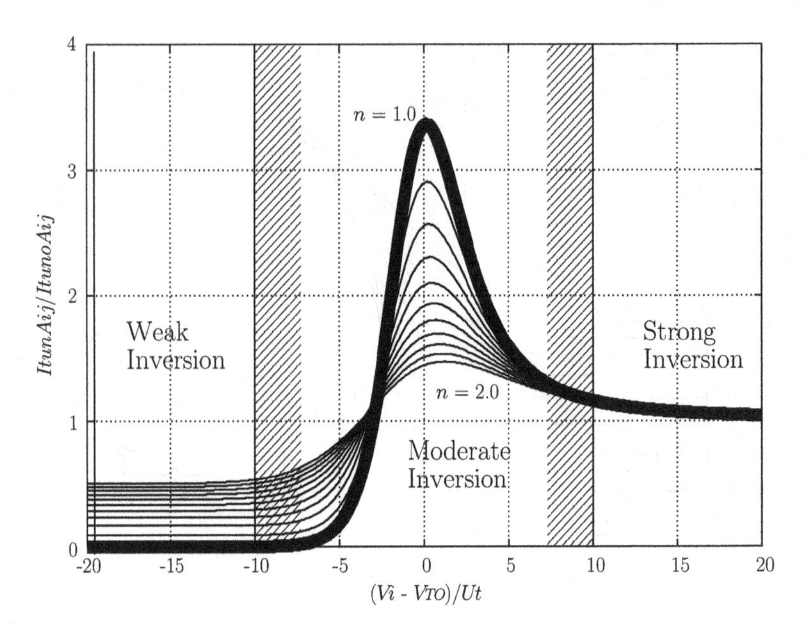

Figure 4.29. Tuning correction factor versus voltage.

Since inner compressed SS variables of the Log filter are limited in practice to a $5U_t$ window, the required transconductance compensation can be simplified. Obviously, the flat regions of C_i in Figure 4.28 simplify such a compensation, unlike around V_{TO} (i.e. moderate inversion). This fact can be easily seen in the required tuning correction of Figure 4.29, computed according to (4.54) for $C_{oi} = C_{oxi}$.

By simple inspection, deep-weak ($IC \ll 1$) and deep-strong ($IC \gg 1$) inversion are the optimum regions of operation for the compressed SS variable V_i. In particular, following the asymptotic values of C_i in (4.56), the tuning compensation is reduced to an almost voltage-independent scaling factor of:

$$I_{tunAij} \simeq \begin{cases} \frac{n-1}{n} I_{tunoAij} & IC \ll 1 \\ I_{tunoAij} & IC \gg 1 \end{cases} \tag{4.58}$$

where the nominal tuning current depends on the MOS compression law used:

$$I_{tunoAij} = \begin{cases} nU_t C_{oxi} A_{ij} & \text{GD} \\ U_t C_{oxi} A_{ij} & \text{SD} \end{cases} \tag{4.59}$$

In principle, both regions of operation (i.e. $IC \ll 1$ or $IC \gg 1$) for the NMOS C_i element can be combined with both companding F functions (i.e. GD or SD) presented in this chapter. However, it seems clear from

Table 4.4. General Filter Performances.

Parameter	Value	Units
Min. Supply Voltage	1.0	V
Technology $(V_{TON} + \|V_{TOP}\|)_{max}$	1.3	V
Full-Scale (I_{max})	4	μA_{pp}
Total Harmonic Distortion @50%I_{max}	< 0.5	%
Dynamic Range @(100Hz-10KHz)	$60 - 70$	dB
Tuning Current (I_{tun})	$1 - 100$	nA
Capacitor/Pole	$10 - 100$	pF
Power/Pole	$5 - 10$	μW
Transistor Area/Pole	0.05	mm^2

the operating point of view that $IC \gg 1$ for C_i is more suitable for GD basic building blocks, while $IC \ll 1$ shows better compatibility with SD cells.

In any case, overall linearity is strongly related to the compressed voltage operating point. In this sense, a brief comparison between these two combinations in Figure 4.29 shows better signal-independence, thus lower distortion, for C_i-SD topologies, while C_i-GD strategies allows larger capacitance-density, thus a more compact Si area. These results can be verified in the last design example of the next section.

7. Design Examples

The purpose of the filter implementations presented in this section is to demonstrate the validity of both new basic building blocks and the design methodology proposed along this chapter.

All examples have been designed for use in very low-voltage audio applications such as CMOS hearing aids systems-on-chip. The common overall performance of the presented implementations is summarized in Table 4.4. The target technologies were both standard 1.2μm and 0.35μm CMOS double-metal double-poly-Si process of Chapter 8. It is foreseen that all designs may be ported to deep sub-micron technologies due to the robustness of the presented circuit techniques. A clear example of this fact is the use of auxiliary ORA blocks to allow minimum channel length in the compressors and expanders of Figure 4.10, thus making the compact area compatible with the wide aspect ratios (typically $(W/L) > 100$) required in these devices to ensure weak inversion operation at full-scale.

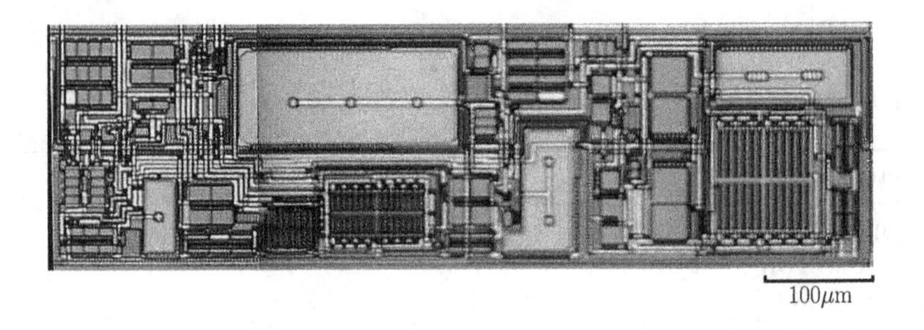

Figure 4.30. Microscope photograph of the saturated SD filter.

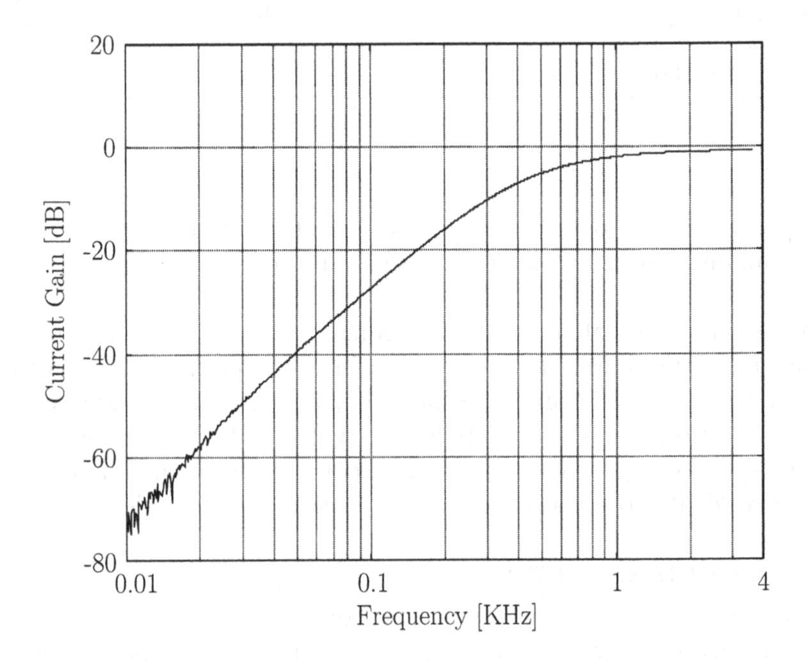

Figure 4.31. Experimental frequency response of the saturated SD filter.

The first circuit realization presented in this section is a one-input one-output second-order high-pass filter implemented through the saturated SD basic building blocks of Figure 4.10. The microscope photograph of the complete integrated filter is depicted in Figure 4.30. Corner frequency is tunable from 100Hz to 10KHz with continuous or digital control. The measured large-signal frequency response at 50% full-scale input for a $f_{-3\text{dB}}$ = 1KHz is represented in Figure 8.10. The filter exhibits an in-band Total Harmonic distortion (THD) of around 0.5% at 50% full-scale output, like that of in Figure 4.32. In this example, a dynamic range (DR) of 63dB is achieved for 10pF per pole.

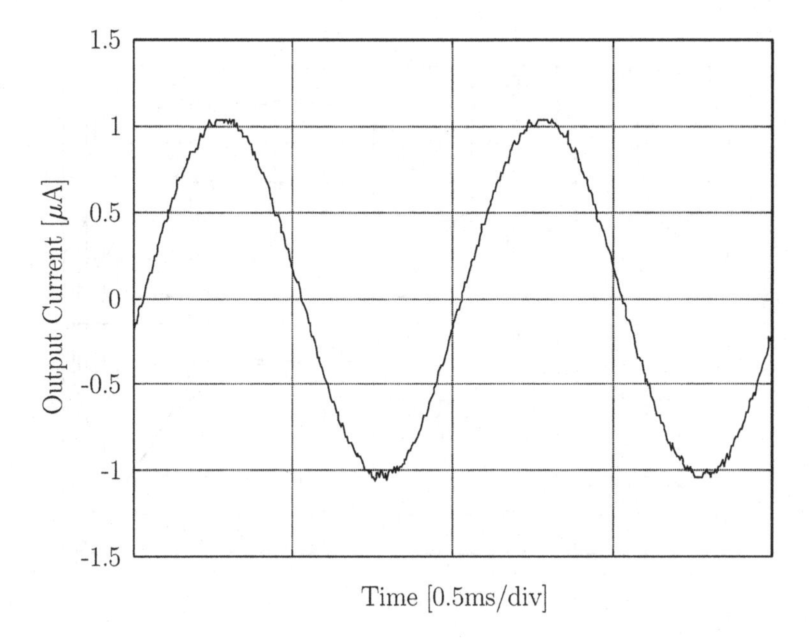

Figure 4.32. Experimental output of the saturated SD filter.

The second design example consists of a one-input two-output second-order band-/low-pass filter, synthesized following Table 4.1 in Table 4.3. Band and low pass responses are available at the first (I_{out1}) and second (I_{out2}) output, respectively. The implementation chosen in this case corresponds to the GD realization of Figure 4.25 with $I_{tuno} = nU_tCw_o$. The tuning capabilities are depicted in Figure 4.33 for different central frequencies and quality factors, showing a typical $THD < 0.4\%$ at a central frequency for 50% full-scale, and $DR = 60$dB for 10pF per pole.

The third example is a modular one-input one-output third-order low-pass filter implemented using both GD and non-saturated SD basic building blocks, as depicted in Figure 4.34. This dual design case has been also integrated with both poly-Si and NMOS capacitors in a standard 0.35μm CMOS technology, as shown in Figure 4.35. An equal-area law has been followed in order to compare results with all-MOS implementations, resulting in $C_{polySi} = 50pF$ and $C_{ox} = 250pF$ due to differences between oxide thickness. Selection of the filter-order can be easily performed here through switch-off and by-pass of cascaded stages. Experimental results are reported in Figures 4.36 and 4.37.

As already pointed, final optimization of signal distortion is strongly related to the compressed voltage operating point, which is set by the level shifter included in both implementations of Figure 4.34 (i.e. the K ratio). The importance of this design parameter can be seen in the

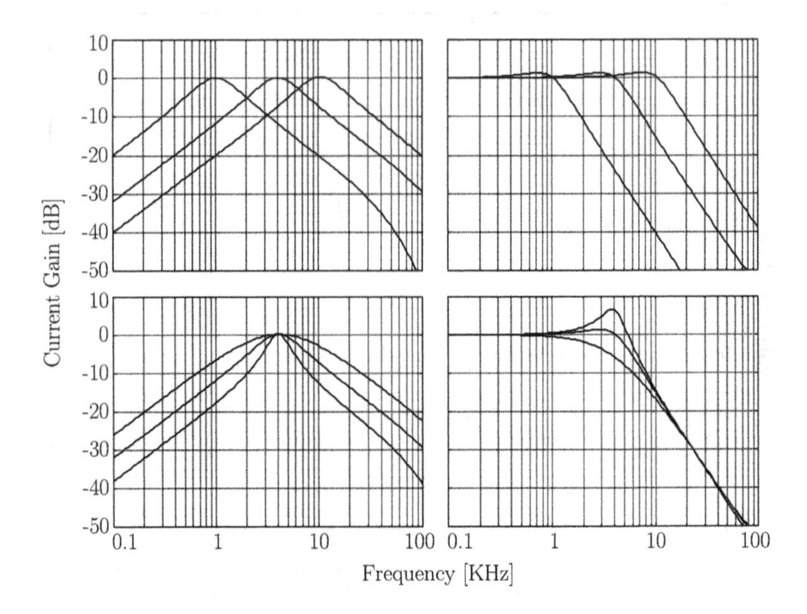

Figure 4.33. Simulated responses of the band (left) and low-pass (right) saturated GD filter for $I_{tuno} = \{2.5\text{nA}, 10\text{nA}, 25\text{nA}\}$ (upper) and $Q = \{1/2, 1, 2\}$ (lower) cases.

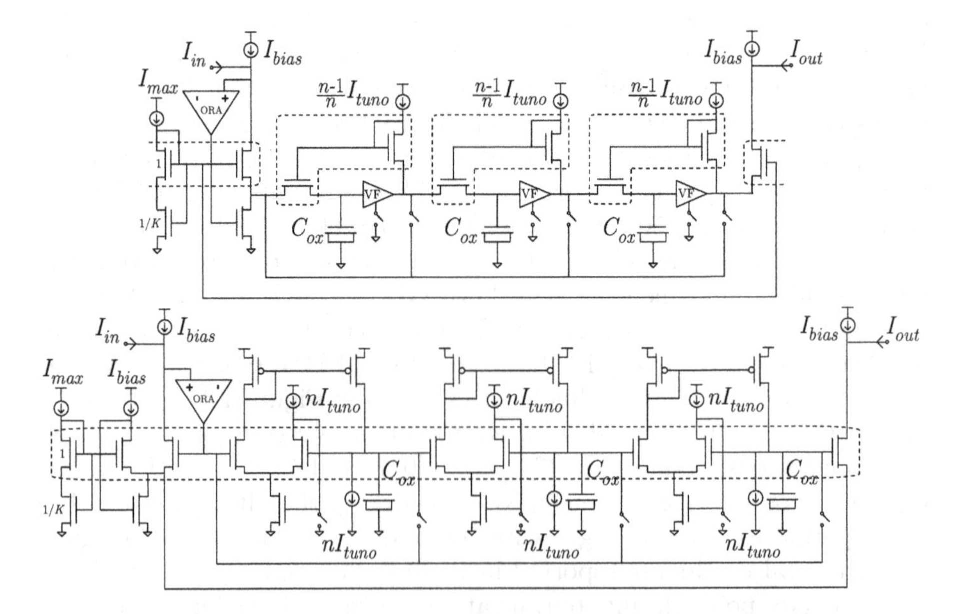

Figure 4.34. All-MOS configurable third-order filter using non-saturated SD (upper) and GD (lower) basic building blocks with $I_{tuno} = U_t C 2\pi f_c$.

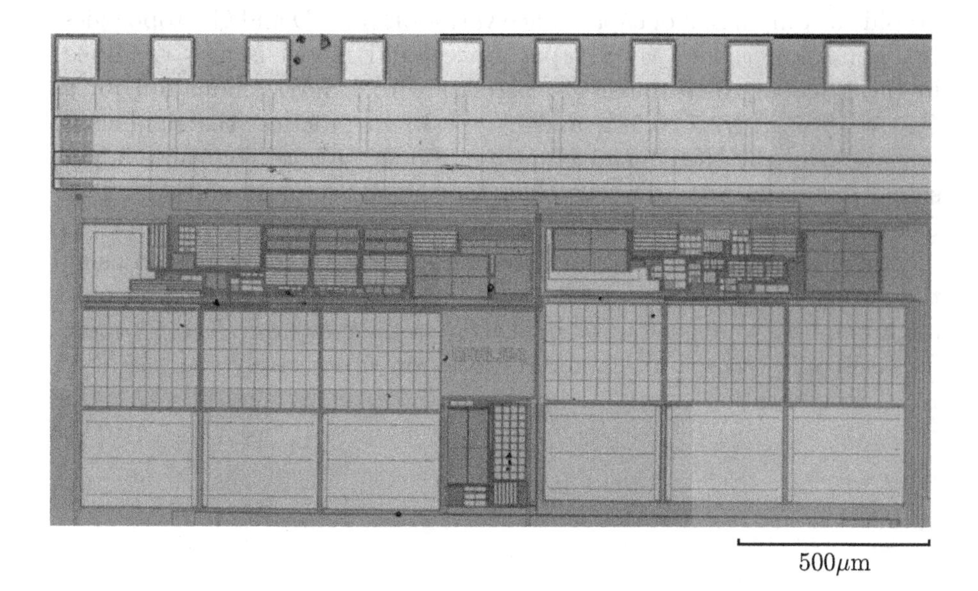

Figure 4.35. Microscope photography of the poly-Si (upper) and all-MOS (lower) non-saturated SD (left) and GD (right) filter in a 0.35μm CMOS technology.

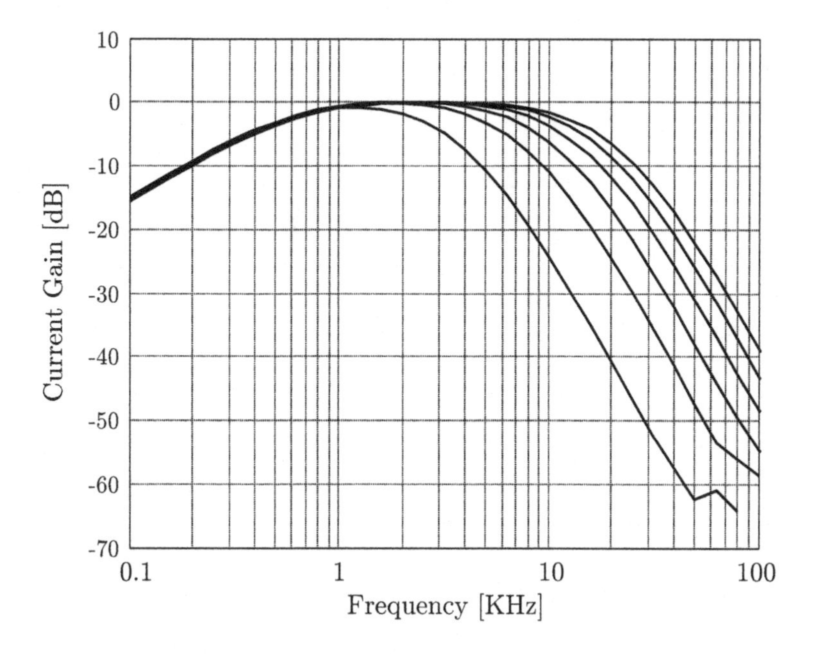

Figure 4.36. Experimental transfer functions at half-full-scale and 3rd-order selection for different tuning currents (input V/I conversion and DC decoupling in test setup implemented using a simple series resistor and capacitor).

graphical comparison of capacitance variations for SD and GD topologies of Figure 4.38. The MOS capacitance hand model is also validated in the same figure through its correlation to BSIM1 simulation for a C_{ox} = 10pF, V_{TO} = 0.54V and n = 1.3. In practice, the significant deviations of the NMOS capacitance in moderate inversion cause larger signal distortion for GD cells than for the SD implementation. This fact can be noted in the THD analysis of Figure 4.37, where distortion increments are about double at near full-scale for GD compression using the same K values of Figure 4.38 and third-order. As a result, while only a minimum $K > 10$ is required in SD compression to ensure enough room for telescopic transistors, larger K ratios or alternatively some kind of I_{tuno} compensation should be investigated.

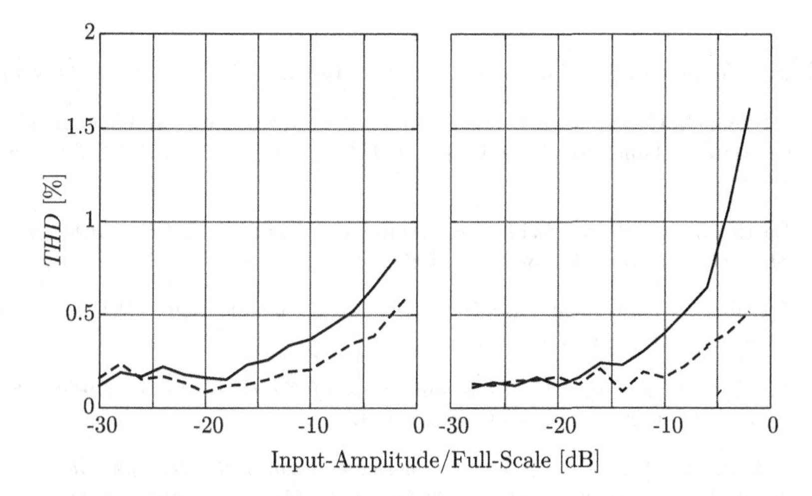

Figure 4.37. Experimental THD of the poly-Si (dashed) and all-MOS (solid) third order non-saturated SD (upper) and GD (lower) circuits versus input signal amplitude.

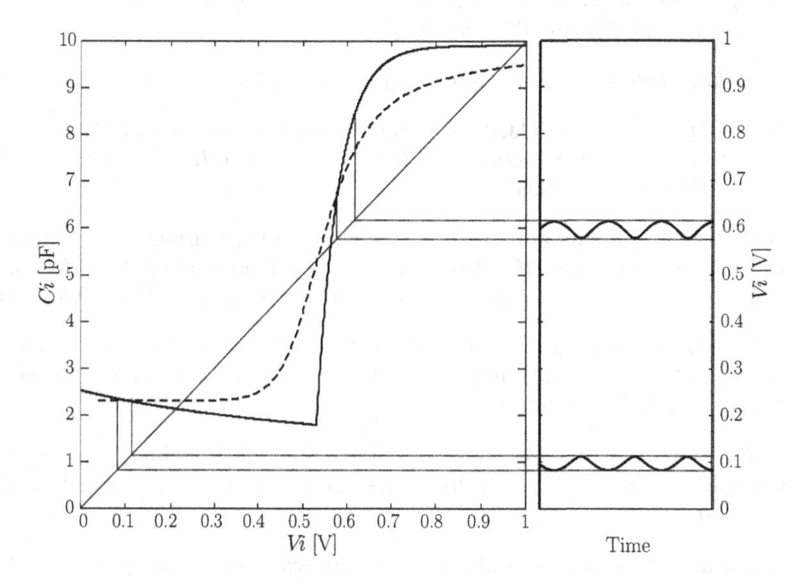

Figure 4.38. Analytical (dashed) and BSIM model (solid) of a $75 \times 75 \mu m^2$ NMOS C (left) at room temperature, and typical compressed signals at 50% full-scale (right) for GD with $K = 80$ (upper), and SD with $K = 20$ (lower).

References

[1] R.W.Adams. Filtering in the Log-Domain. In *63rd AES Conference*, May 1979.

[2] E.Seevinck. Companding Current-Mode Integrator: A New Circuit Principle for Continuous-Time Monolithic Filters. *IEE Electronics Letters*, 26(24):2046–2047, 1990.

[3] Barrie Gilbert. A New Wide-Band Amplifier Technique. *IEEE Journal of Solid State Circuits*, 3(4):353–365, Dec 1968.

[4] B.Gilbert. Translinear Circuits: a Proposed Classification. *IEE Electronics Letters*, 11(1):14–16, Jan 1975.

[5] E.A.Vittoz. *Analog VLSI Implementation of Neural Networks*, chapter E1.3, pages 1–17. Oxford University Press, 1997.

[6] B.A.Minch. *Analysis, Synthesis, and Implementation of Networks of Multiple-Input Translinear Elements.* PhD thesis, California Institute of Technology, Pasadena, California, May 1997. http://www.ee.cornell.edu/~minch/.

[7] T.Serrano-Gotarredona, B.Linares-Barranco, and A.G.Andreou. A General Translinear Principle for Subthreshold MOS Transistors. *IEEE Transactions on Circuits and Systems-I*, 46(5):607–616, May 1999.

[8] D.R.Frey. Log-Domain Filtering: an Approach to Current-Mode Filtering. *IEE Proceedings*, 140(6):406–415, Dec 1993.

[9] K.Ogata. *Modern Control Engineering.* Prentice-Hall Inc., 1970.

[10] R.Fried, D.Python, and C.C.Enz. Compact Log-Domain Current Mode Integrator with High Transconductance-to-Bias Current Ratio. *IEE Electronics Letters*, 32(11):952–953, May 1996.

[11] D.Python, R.Fired, and C.C.Enz. A 1.2V Companding Current-Mode Integrator for Standard Digital CMOS Processes. In *Third International Conference on Electronics, Circuits and Systems*, volume 1, pages 231–234. IEEE, Oct 1996.

[12] C.Enz, M.Punzenberger, and D.Python. Low-Voltage Log-Domain Signal Processing in CMOS and BiCMOS. *IEEE Transactions on Circuits and Systems-II*, 46(3):279–289, Mar 1999.

[13] N.Krishnapura, Y.Tsividis, and D.R.Frey. Simplified Technique for Syllabic Companding in Log-Domain Filters. *IEE Electronics Letters*, 36(15):1257–1259, Jul 2000.

[14] W.Chen. *Linear Networks and Systems: Algorithms and Computer-Aided Implementations*, volume 3 of *Advanced Series in Electrical and Computer Engineering.* World Scientific, second edition, 1990.

Chapter 5

PTAT GENERATION

Abstract This chapter introduces new circuit techniques devoted to the generation of static I/V PTAT references. The topologies obtained can be understood as a particularization of the general design techniques proposed for amplification in Chapter 3. An integrated example is also presented as a demonstrator of the novel Log companding strategy.

1. Log Companding Principle

As already pointed out in the previous chapters, I/V static references proportional-to-absolute-temperature (PTAT) are needed for most of the proposed circuit techniques in order to cancel first-order thermal dependencies on tuning. In particular, the following magnitudes are mandatory:

- PTAT voltage references for gain control according to (3.9).
- PTAT current references for filter tuning from (4.29).

Furthermore, such generators should exhibit the best low-voltage compatibility as they supply reference levels for other parts of the system. In this sense, previously reported PTAT generators require bipolar devices and resistors [1, 2, 3] or do not exhibit enough low-voltage capabilities for sub-1V operation [4, 5, 6].

The new design approach makes use of Log companding processing to allow strong supply scaling. In fact, from the general device-independent y/x nomenclature introduced in Chapter 1, we already know that $x \propto V/U_t$, so obtaining a voltage PTAT reference is equivalent to synthesizing a constant value in the compressed domain. The basic idea consists of describing the PTAT voltage generator as the Log amplifier (G) defined in Subsection 3.1 within a fixed attenuation feedback

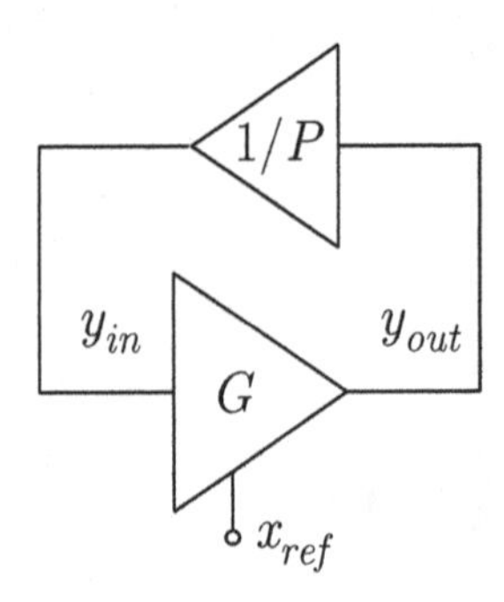

Figure 5.1. Log companding proposal for the PTAT generator.

$(1/P)$ as depicted in Figure 5.1. Due to the feedback loop, the controllable amplifier operates at forced input and output and the control port (x_{ref}) becomes the effective result.

From its original expression (3.3), the controllable Log amplifier exhibits the following gain:

$$G = \frac{y_{out}}{y_{in}} \doteq e^{x_{ref}} \tag{5.1}$$

Feedback in Figure 5.1 sets $GP \equiv 1$, causing the control terminals of the amplifier to exhibit:

$$x_{ref} = \ln P \tag{5.2}$$

so the desired PTAT reference V_{ref} is obtained due to the normalizing factor U_t in (1.3), while low-voltage operation is achieved by its Log compression with respect to circuit currents. The corresponding current reference (I_{ref}) can be easily synthesized through an attached impedance at the control port V_{ref}. Linearity of this load element is only necessary in case of PTAT specifications for I_{ref}, too.

An important design parameter of any reference is its accuracy. In our case, the main source of uncertainty in equation (5.2) comes from the resolution of the P factor. Hence, it is convenient to express the relative accuracy on x_{ref} in terms of:

$$\left(\frac{\Delta x_{ref}}{x_{ref}}\right) = \frac{\ln\left(1 + \frac{\Delta P}{P}\right)}{\ln P} \simeq \frac{1}{\ln P}\left(\frac{\Delta P}{P}\right) \qquad \Delta P \ll P \tag{5.3}$$

Due to the log dependence on P, maximum x_{ref} sensitivity occurs at $P \to 1^+$, while x_{ref} robustness increases for $P \to \infty$. Hence, high sensitivity should be avoided in favor of maximum P ratios, thus larger

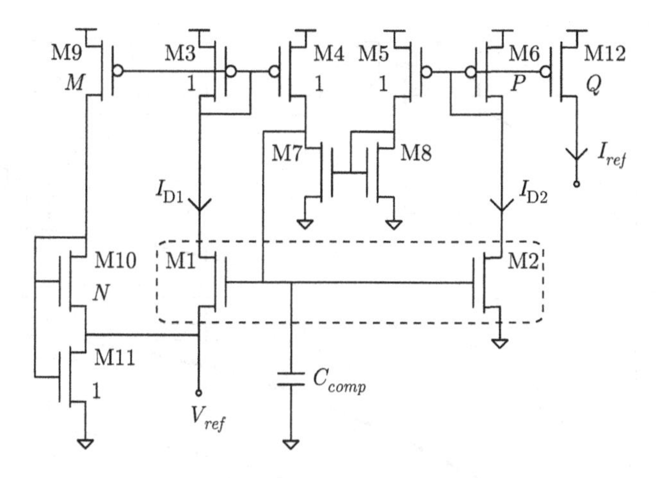

Figure 5.2. Schematic of the proposed PTAT generator.

x_{ref} values. In CMOS implementations of Figure 5.1, ΔP is typically associated to technology mismatching at transistor level, and must be taken into account during the design process as discussed in the next subsection.

2. CMOS Generalization

From the point of view of the circuit, fixed feedback $1/P$ can be obtained through simple geometry scaling (i.e. current mirrors). On the other hand, the logarithmic gain G will be synthesized here through the GD-SC cell developed in Chapter 3 and depicted in Figure 3.3. Following this approach, the complete schematic of the proposed low-voltage CMOS PTAT generator is depicted in Figure 5.2. All-MOS devices in the schematic are biased at strong inversion saturation, except for M1-M2 and M11 which are operated at weak inversion saturation and strong inversion conduction, respectively.

In this case, the Log amplifier cell is configured by $V_{gaini} \equiv V_{ref}$ and $V_{gaino} \equiv 0$, so that the resulting gain expression from (3.9) is:

$$G = \frac{I_{out}}{I_{in}} \doteq e^{\frac{V_{ref}}{U_t}} \tag{5.4}$$

thus:

$$V_{ref} = U_t \ln P \tag{5.5}$$

The role of the operational transresistance amplifier (ORA) composed of M3-M8 is to fix the feedback current factor $1/P$, as well as a low-enough impedance for both the input and output ports of the G stage,

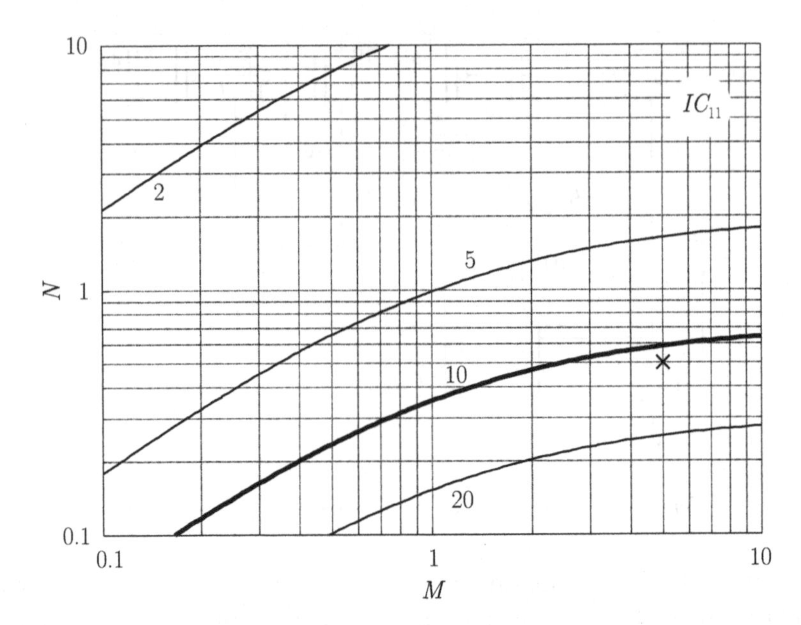

Figure 5.3. M11 inversion coefficient (IC_{11}) versus M and N design space for $P = 10$. The cross-point indicates the design example of Section 3.

following the same technique of Figure 3.6. Because of the latter, the resulting cancellation of channel length modulation (CLM) effects in M1-M2 allows minimum L lengths to compact the Si area of these devices.

Once the PTAT voltage reference has been synthesized, the non-linear MOS impedance M9-M11 is attached to the PTAT voltage core. The resulting copy of the current reference through M12 follows as:

$$I_{ref} = Q \left[\frac{\ln P}{2(M+1)} \left(\sqrt{\frac{M}{N}} + \sqrt{\frac{M}{N} + M + 1} \right) \right]^2 I_{S11} \qquad (5.6)$$

which is valid as long as M10-M11 are kept in strong inversion operation (i.e. $\frac{(M+1)}{Q} I_{ref} \gg I_{S11}$). In this sense, the complete M-N design space is explored in Figure 5.3. The strong inversion specification for M11 causes all suitable solutions to be located more than one decade above the specific current (i.e. $IC_{11} > 10$), as indicated in the same figure.

In fact, the absolute value of I_{ref} is obtained from the specific current itself, the only technological current magnitude available at device level. Hence, the current reference must be designed by choosing a suitable $(W/L)_{11}$ to scale the NMOS unitary I_S (I_{SNu}) of the CMOS process. The main drawback of this MOS implementation is its technology sensitivity to β and its non-PTAT behaviour (i.e. $\propto \beta U_t^2$). On the other hand, such I_S dependence is desired for the current bias according to (7.5) in order to keep signal distortion constant against temperature.

As pointed out in the previous section, V_{ref} accuracy is only dependent on P resolution, so two key points must be considered in the loop gain of Figure 5.2: deviations due to channel length modulation in the M1-M2 pair, and technological mismatching at the device level. The first non-ideal effect is compensated by the ORA block, which keeps a balance between V_{DB1} and V_{DB2} voltages. Concerning device mismatching, two simplifications will be taken following the technology mismatch model of Section 5: at similar device area, relative drain current variations in transistors operating in strong inversion are negligible compared to those which are biased in subthreshold; and threshold voltage mismatching (ΔV_{TO}) is dominant over current factor one ($\Delta\beta$). Accepting both hypothesis, relative deviations of the P factor can be expressed from (2.28) as:

$$\sigma\left(\frac{\Delta P}{P}\right) \simeq \frac{\sigma\left(\Delta V_{TO}\right)}{nU_t} = \frac{1}{\sqrt{(WL)_{1,2}}}\left(\frac{A_{VTO}}{nU_t}\right) \tag{5.7}$$

where A_{VTO} stands for the technological mismatching parameter of V_{TO}. Typically, $A_{VTO} \simeq 15\text{mV}\mu\text{m}$ for $1\mu\text{m}$ CMOS technologies as depicted in Figure 2.8. Thus, after combining (5.3), (5.5) and (5.7), the final resolution of V_{ref} can be directly related to the Si area according to:

$$\sigma\left(\frac{\Delta V_{ref}}{V_{ref}}\right) = \frac{1}{\sqrt{(WL)_{1,2}}}\left(\frac{A_{VTO}}{nV_{ref}}\right) \tag{5.8}$$

3. Design Examples

In the first example, initial specifications are given in terms of absolute reference values $V_{ref} \simeq 60\text{mV}$ and $I_{ref} \simeq 320\text{nA}$, and relative deviations $(\Delta V_{ref}/V_{ref}) \leq 5\%$ all at room temperature. Design flow starts by computing the required feedback factor for the PTAT voltage core to achieve the desired V_{ref} according to (5.5), resulting $P = 10$. Once V_{ref} is fixed, a resistor $R_{ref} = 390\text{K}\Omega$ has been chosen here to generate the required PTAT I_{ref} for its application in CMOS Log-domain filter-tuning [1]. The accuracy specification of V_{ref} is translated into an equivalent standard deviation using a 2σ law (i.e. 96% of samples), resulting $\sigma\left(\Delta V_{ref}/V_{ref}\right) \leq 2.5\%$. Hence, the required Si area for the M1-M2 pair $(WL)_{1,2} \geq 85\mu\text{m}^2$ is obtained from (5.8) and also from parameters of the target process $A_{VTO} = 18\text{mV}\mu\text{m}$ and $n = 1.3$. The resulting CMOS

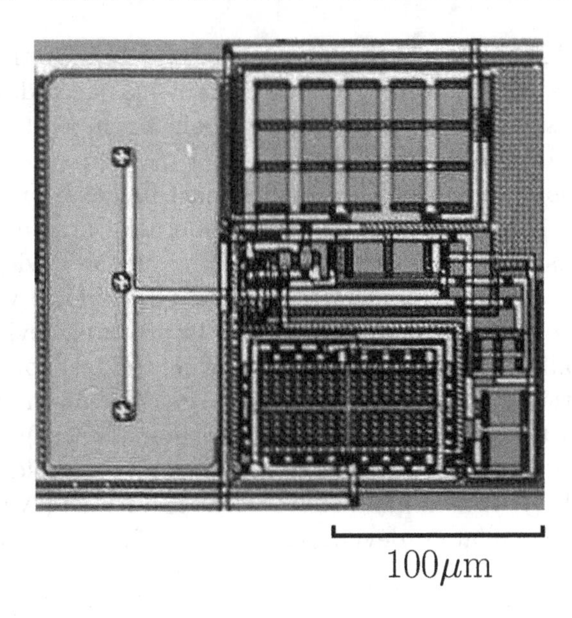

Figure 5.4. Microscope photography of the 1.2μm PTAT generator.

Table 5.1. Summary of typical results for the PTAT generator.

Parameter	Simulated	Experimental	Units
Min. Supply Voltage	0.9	0.9	V
Technology $(V_{TON}+\|V_{TOP}\|)_{max}$	1.3	not available	V
Nominal V_{ref}	62	65	mV
$2\sigma(\Delta V_{ref}/V_{ref})$	5.5	5.4	%
V_{ref} Sensitivity to T	+0.22	+0.22	mV/°C
$PSRR+(DC)$	>70	>60	dB
Nominal I_{ref}	320	345	nA
$2\sigma(\Delta I_{ref}/I_{ref})$	5.5	6	%
I_{ref} Sensitivity to T	+1.15	+1.13	nA/°C
Power Consumption	2.5	not available	μW
Si Area	0.05	0.05	mm^2

PTAT circuit has been integrated using a 1.2μm double-metal double-poly-Si CMOS technology, exhibiting a total Si area of about 0.05mm^2 as shown in Figure 5.4. A comparison between experimental, BSIM3 simulated and analytical results is reported in Table 5.1 and Figures 5.5 to 5.7. A $PSRR(DC)+ > 40$dB criteria has been chosen to define the minimum supply voltage of the generator, which exhibits sub-1V operation capabilities.

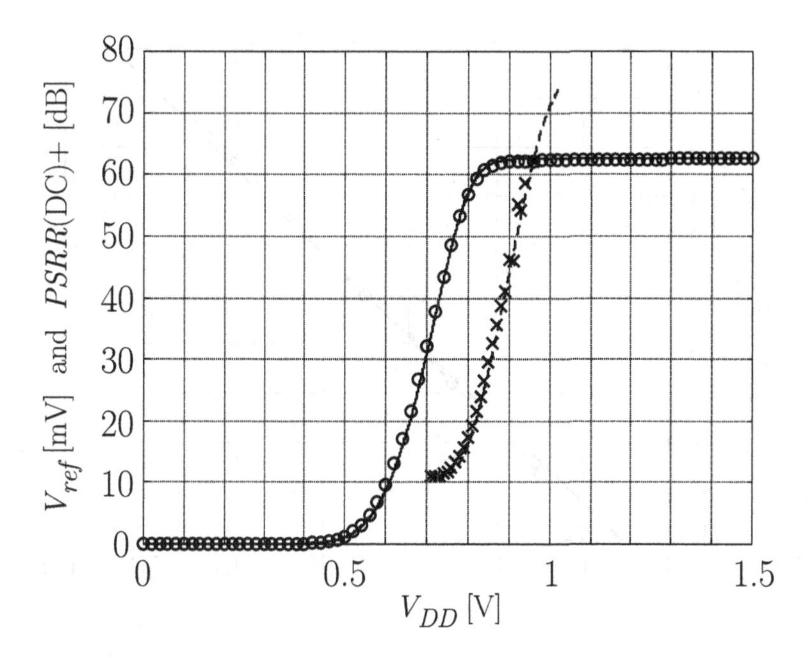

Figure 5.5. Experimental (dotted and crossed) and simulated (solid and dashed) V_{ref} and static $PSRR+$ respectively versus supply voltage at room temperature for the 1.2μm PTAT generator.

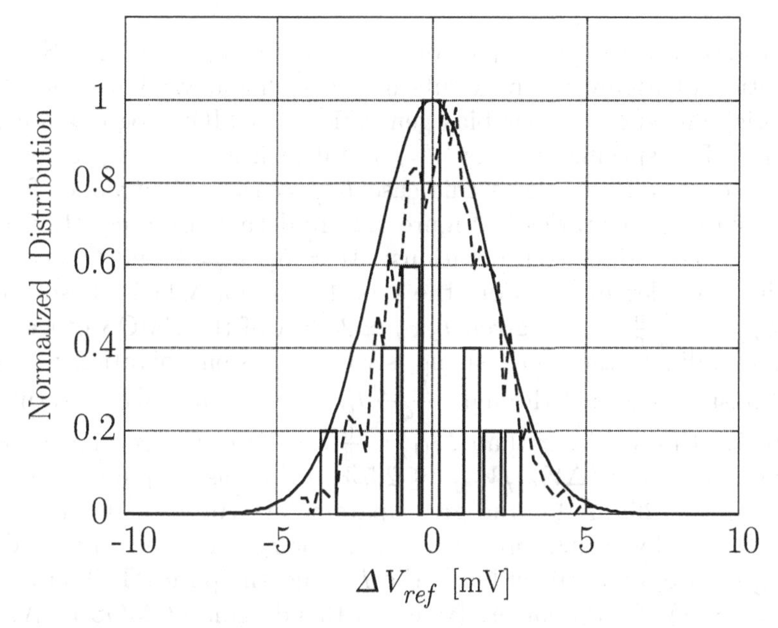

Figure 5.6. Experimental (bar), simulated (dashed) and analytical (solid) V_{ref} histogram at room temperature from 15 samples and 1000 Montecarlo runs for the 1.2μm PTAT generator.

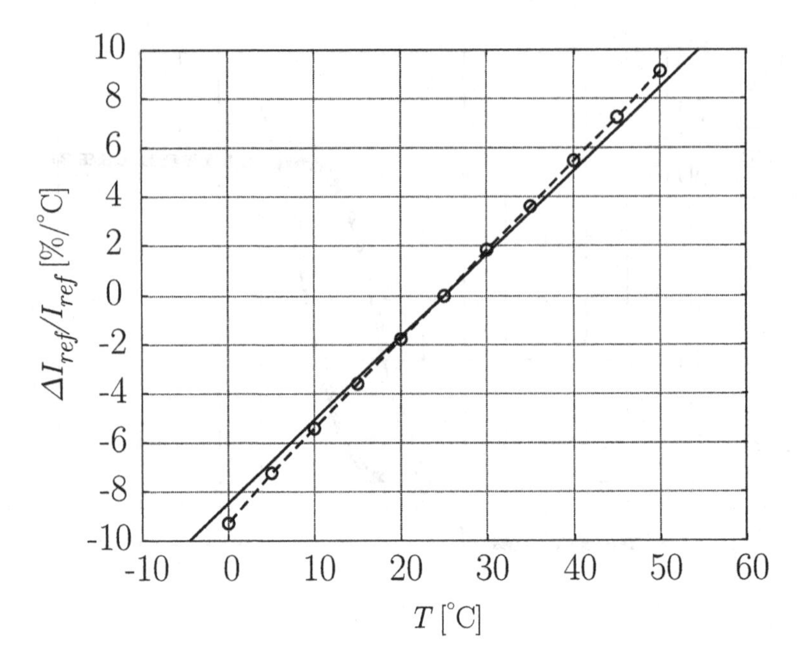

Figure 5.7. Experimental (dotted), simulated (dashed) and analytical (solid) relative I_{ref} variations versus temperature for the 1.2μm PTAT generator.

The second circuit example consists of the complete all-MOS implementation of Figure 5.2 in a sub-micron technology. This case study was originally designed for bias generation in CMOS Log domain processing. The specification for V_{ref} is taken from to the previous example, so $P = 10$, while in this case $I_{ref} = 1\mu A$. The values for the rest of inter-device ratios in Figure 5.2 are derived from equation (5.6) and also from power considerations: $M = 5$, $N = \frac{1}{2}$ and $Q = 3$. Using the same design equation, the aspect ratio for M11 is chosen to be $(W/L)_{11} = \frac{10\mu m}{16\mu m}$ for the given $I_{SNu} \simeq 280nA$ of the CMOS process. In order to validate the above sizing, strong inversion operation for M10-M11 must be checked through $\frac{(M+1)}{Q} I_{ref} \equiv \frac{6}{3} \times 1\mu A$, which results in about five times greater than $I_{S11} = \frac{10}{16} \times 280nA$. Concerning V_{ref} accuracy, the same $\sigma\left(\Delta V_{ref}/V_{ref}\right) \leq 2.5\%$ in this case requires from (5.8) a $(WL)_{1,2} \geq 26\mu m^2$ for the given parameters $A_{V_{TO}} = 10mV\mu m$ and $n = 1.3$. Finally, a dual operation mode for I_{ref} has been included following Figure 5.8 to allow both, the I_S- and the pure PTAT-law, with $R_{ref} = 180K\Omega$ for the latter. Based on this design, a CMOS PTAT circuit has been integrated in a 0.35μm digital VLSI technology as shown in Figure 5.9. Since V_{ref} behaviour is similar to first example, only the comparative results for I_{ref} are presented in Figures 5.10 and 5.11. Both

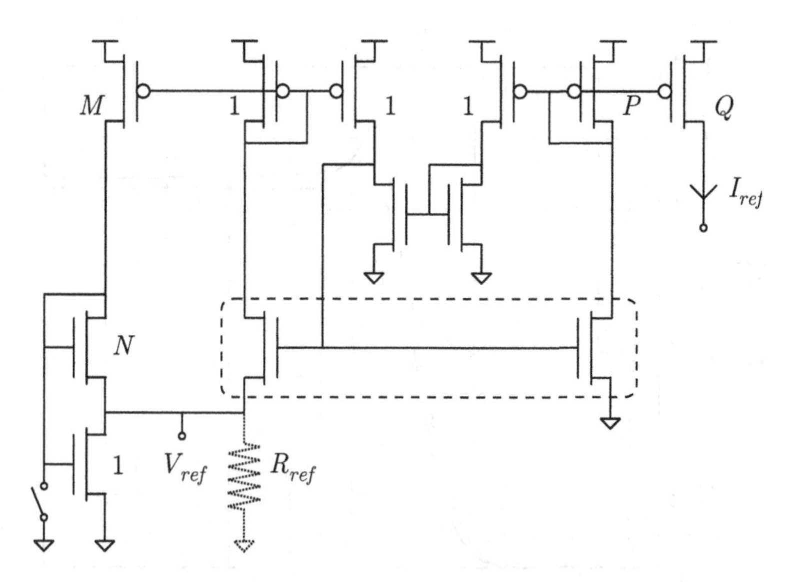

Figure 5.8. Schematic of the V_{ref} PTAT generator with dual-mode for I_{ref}: I_S-based (switch-off and no-R_{ref}) or PTAT (switch-on and R_{ref}).

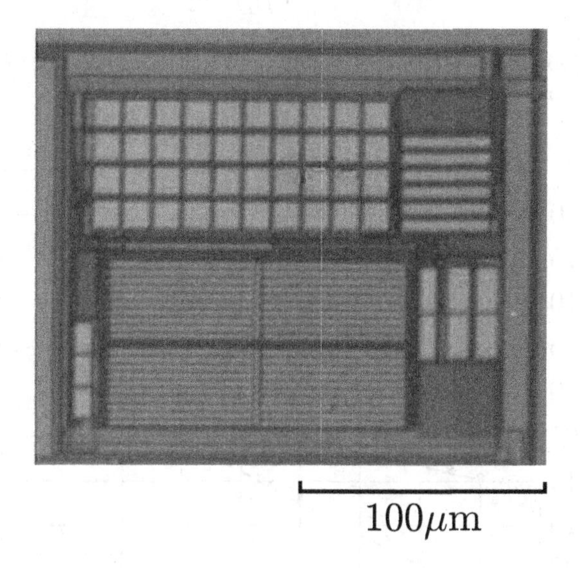

Figure 5.9. Microscope photography of the 0.35μm PTAT generator.

modes of operation exhibit almost the same relative current deviations $\sigma(\Delta I_{ref})/I_{ref}$, ranging from 2.5% to 3.5%.

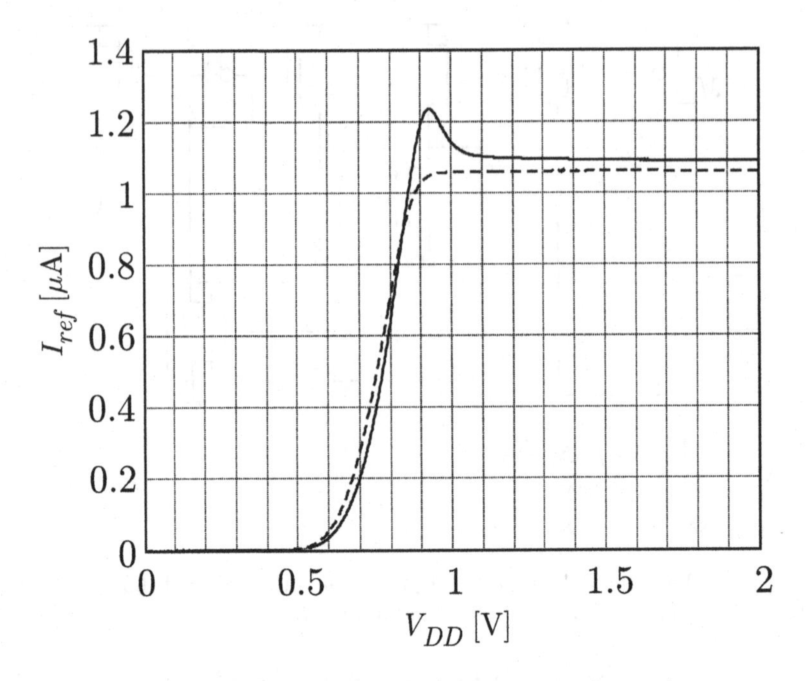

Figure 5.10. Experimental all-MOS (solid) and R_{ref}-based (dashed) I_{ref} versus supply voltage at room temperature for the 0.35μm PTAT generator.

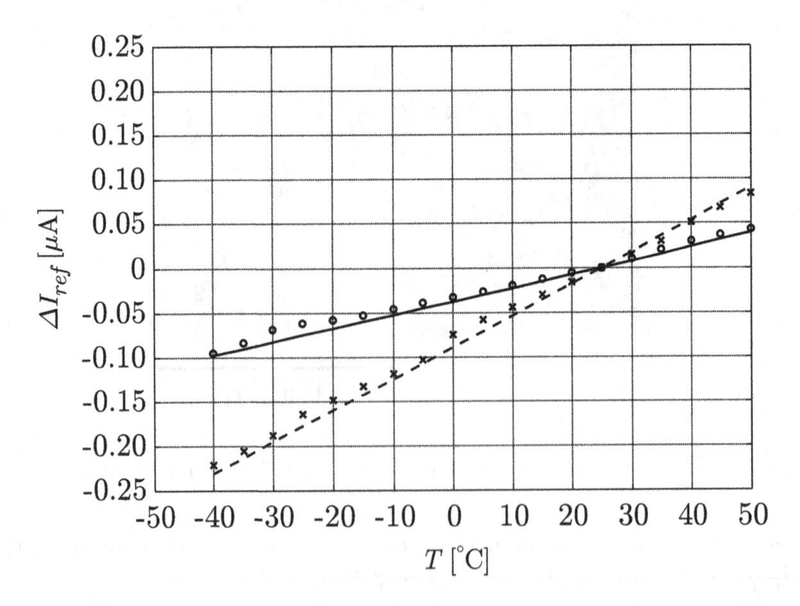

Figure 5.11. Experimental (dots) and simulated (solid) all-MOS, and experimental (crosses) and simulated (dashed) R_{ref}-based I_{ref} variations versus temperature for the 0.35μm PTAT generator.

References

[1] Y.Jiang and E.K.F.Lee. Design of Low-Voltage Bandgap Reference Using Transimpedance Amplifier. *IEEE Transactions on Circuits and Systems-II*, 47(6):552–555, Jun 2000.

[2] H.Banba, H.Shiga, A.Umezawa, T.Miyaba, T.Tanzawa, S.Atsumi, and K.Sakui. A CMOS Bandgap Reference Circuit with Sub-1-V Operation. *IEEE Journal of Solid State Circuits*, 34(3):670–674, May 1999.

[3] Y-D.Seo, D.Nam, B-J.Yoon, I-H.Choi, and B.Kim. Low-Power CMOS On-Chip Voltage Reference Using MOS PTAT: An EP Approach. In *Proceedings of the ASIC Conference and Exhibit*, pages 316–320. IEEE, Sep 1997.

[4] A.Buck, C.McDonald, S.Lewis, and T.R.Viswanathan. A CMOS Bandgap Reference without Resistors. In *Proceedings of the International Solid-State Circuits Conference*, pages 442–443. IEEE, 2000.

[5] H.J.Oguey and D.Aebischer. CMOS Current Reference without Resistance. *IEEE Journal of Solid State Circuits*, 32(7):1132–1135, Jul 1997.

[6] H.Sanchez, R.Philip, J.Alvarez, and G.Gerosa. A CMOS Temperature Sensor for $PowerPC^{TM}$ RISC Microprocessors. In *Proceedings of the Symposium on VLSI Circuits*, pages 13–14. IEEE, Jun 1997.

References

Chapter 6

PULSE DURATION MODULATION

Abstract This chapter focuses on the application of the filtering techniques presented in Chapter 4 to signal modulation in the Log domain. The research includes all the new basic building blocks used to synthesizing fully integrable pulse duration modulators (PDM). An experimental example is also supplied at the end to verify the validity of the proposed circuit technique.

1. Log Companding Principle

Pulse-duration modulation (PDM) processing is of special interest in output power stages for driving low-impedance loads in Class-D [1]. The main advantages of its use are optimization of both the available output signal range and the quiescent power consumption. In particular, portable low-power CMOS system-on-chip applications like hearing aids require this type of signal modulation in order to extend battery life.

Unfortunately, most previous CMOS solutions do not provide a low-voltage compatibility good enough for true single-battery cell operation (down to 1V) [2, 3, 9], thus requiring supply multipliers which tend to decrease power efficiency, increase external components and Si area overhead as explained in Chapter 1. Other reported proposals make extensive use of resistors or bipolar devices [5, 6]. A novel very low-voltage CMOS circuit strategy for PDM is presented here in the Log companding frame of this study.

In general, a PDM signal (y_{out}) can be obtained from a 1-bit comparison between the base-band input (y_{in}) and a higher frequency triangle waveform (y_{tri}) as depicted in Figure 6.1. The signal y_{tri} may be generated by integrating a constant (y_{ref}), which is periodically inverted according to a memory element (e.g. flip-flop) controlled by the output window (y_{thmin},y_{thmax}).

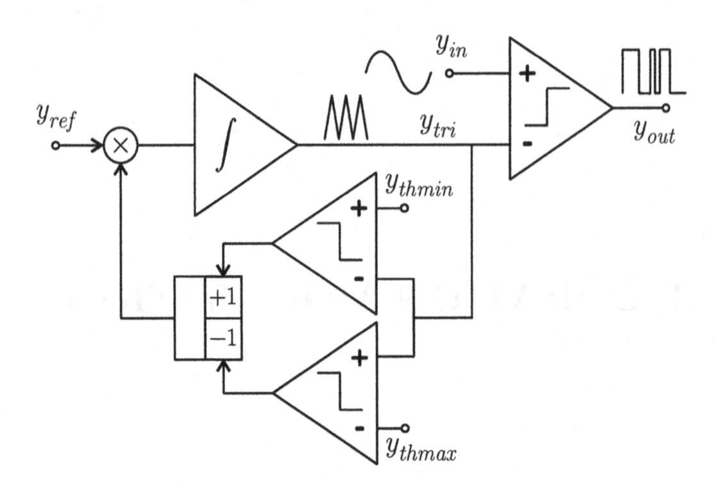

Figure 6.1. General design approach for PDM generation.

From the point of view of signal processing:

$$\frac{\mathrm{d}y_{tri}}{\mathrm{d}t} = \pm\frac{y_{ref}}{\tau} \qquad y_{thmin} < y_{tri} < y_{thmax} \tag{6.1}$$

where τ stands for the integrator time constant. The triangle waveform generated at y_{tri} exhibits a periodicity of:

$$T_{tri} = 2\frac{y_{thmax} - y_{thmin}}{y_{ref}} \tag{6.2}$$

Taking into account the Log companding environment of this study, the comparison and windowing process from Figure 6.1 can be easily synthesized in the linear y-domain (i.e. current domain) through simple KCL algebra. Hence, the main design problem is related to the differential equation of the integrator which generates y_{tri} in (6.1). In this sense, the required processing in the compressed x-domain is obtained by applying the Log companding function F from (1.5) to the integrator equation (6.1):

$$\frac{\mathrm{d}x_{tri}}{\mathrm{d}t} = \pm\frac{1}{\tau}e^{x_{ref}-x_{tri}} \tag{6.3}$$

As explained in Section 1, the above expression is synthesized physically by storing the state-space (SS) variable x_{tri} across a capacitor C:

$$C\frac{\mathrm{d}x_{tri}}{\mathrm{d}t}\frac{U_t}{I_S} = \pm y_{tun}e^{x_{ref}-x_{tri}} \qquad y_{tun} \doteq \frac{CU_t}{\tau I_S} \tag{6.4}$$

where y_{cap} can be identified as the charge and discharge current while y_{tun} corresponds to the tuning parameter.

2. CMOS Generalization

Coming back to the CMOS implementation, the device-independent design equation (6.1) is associated with the external I-domain as:

$$\frac{\mathrm{d}I_{tri}}{\mathrm{d}t} = \pm\frac{I_{ref}}{\tau} \qquad I_{thmin} < I_{tri} < I_{thmax} \tag{6.5}$$

At this point, the MOS Log companding function F must be chosen between the gate-driven (GD) and source-driven (SD) alternatives proposed in Chapter 4. In this case, the GD option is selected since it returns in general more compact circuit realizations:

$$I = F(V) = I_S e^{-\frac{V_{TO}+nV_{bias}}{nU_t}} e^{\frac{V}{nU_t}} \qquad I_S = 2n\beta U_t^2 \tag{6.6}$$

Now, applying the above $I \rightarrow V$ compression law to (6.5) yields:

$$\frac{\mathrm{d}V_{tri}}{\mathrm{d}t} = \pm\frac{nU_t}{\tau} e^{\frac{V_{ref}-V_{tri}}{nU_t}} \tag{6.7}$$

Again, in case the compressed signal V_{tri} is stored across a grounded capacitor (C), the expression in (6.7) can be decribed in the charge domain (Q) as a non-linear transconductance driving of C:

$$\frac{\mathrm{d}Q_{tri}}{\mathrm{d}t} = \underbrace{C\frac{\mathrm{d}V_{tri}}{\mathrm{d}t}}_{I_{cap}} = \pm I_{tun} e^{\frac{V_{ref}-V_{tri}}{nU_t}} \qquad I_{tun} \doteq \frac{nU_tC}{\tau} \tag{6.8}$$

where I_{tun} stands for the tuning current. Consequently, the period of the triangular oscillator is defined by:

$$T_{tri} = 2\frac{nU_tC}{I_{tun}}\left(\frac{I_{thmax} - I_{thmin}}{I_{ref}}\right) \tag{6.9}$$

with I_{thmax} and I_{thmin} being the corresponding window boundaries included in Figure 6.1.

The general filtering techniques developed in Chapter 4 can be applied here to implement such an integrator. However, in the particular case of a constant and switched input, a specific synthesis may return an even more compact CMOS realization. By inspection of (6.8), an optimized implementation is proposed in Figure 6.2. The triangle generator is displayed at the top of the figure and built by transistors M1 to M18. As may be seen, since the compressor M1 is biased to a fixed input level I_{ref}, it can be strongly simplified, requiring only M11 to

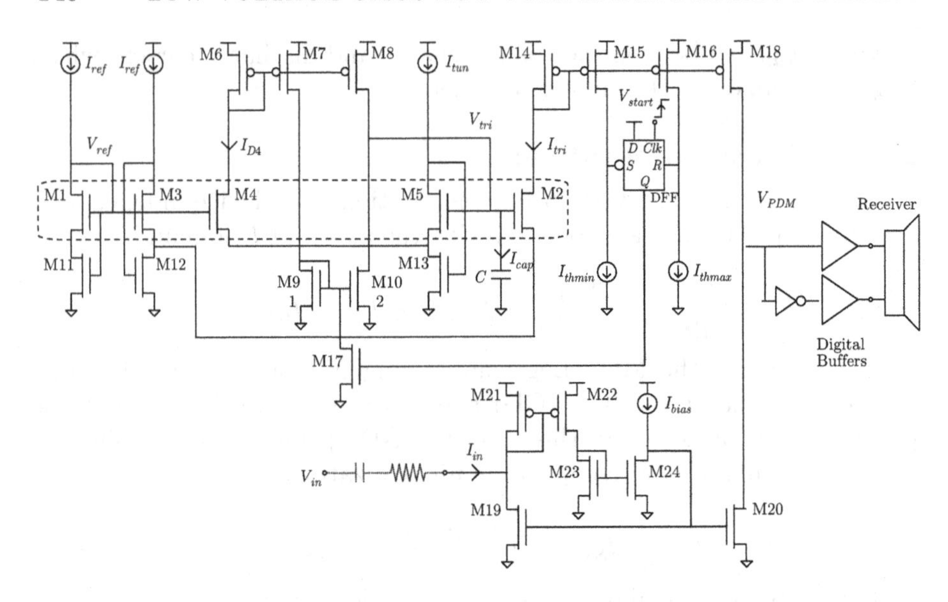

Figure 6.2. Simplified schematic of the PDM modulator.

ensure proper operation of all telescopic devices (i.e. M12 and M13). Also, instead of changing the input value $\pm I_{ref}$ according to Figure 6.1, the sign of the non-linear transconductance M4-M5 is switched through M6-M10, so that $I_{cap} = \pm I_{D4}$. As a result, charge and discharge phases of the integrator capacitor C can be controlled by the NMOS switch M17. The expansion process of the compressed triangle signal V_{tri} is performed through M2, which is attached to the common reference of the compressor thanks to the low-impedance source M3-M12. Oscillation is guaranteed by the feedback window comparator M15-M16, which alternatively changes the sign of the slope stored in a D-type flip-flop (DFF). Finally, some start-up signal (V_{start}) may be required at the flip-flop element for a proper initialization of the cycle.

An example of the resulting Log compressed V_{tri} and expanded I_{tri} triangle waveforms is displayed in Figure 6.3. Again, low-voltage capabilities due to internal compression may be seen clearly in the voltage signal of the same figure. An adapter made of devices M19 to M24 is also included in Figure 6.2 to ensure a low-enough input impedance. Such a feature is of special interest when requiring an additional linear $V \rightarrow I$ conversion of the incoming signal. In these cases, a simple series resistor and eventually a DC decoupling capacitor can be used as indicated in Figure 6.2. The PDM output signal (V_{PDM}) is computed from the 1-bit voltage quantization of $(I_{tri} - I_{in})$. A proper quiescent bias $I_{bias} = (I_{thmax} + I_{thmin})/2$ must be designed to obtain a 50% PDM

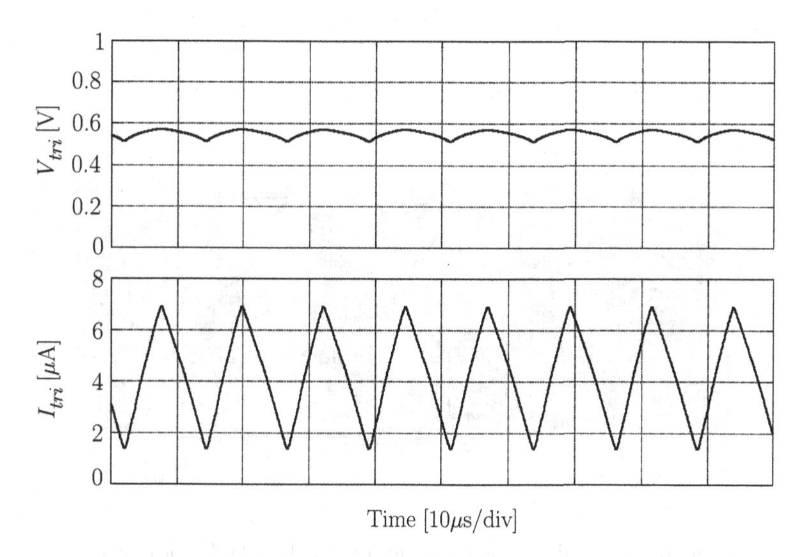

Figure 6.3. Simulated voltage across the integrating capacitor (upper) and output current (lower) when operating at 1.0V voltage supply.

output in silence. Also, typical load conditions for audio applications are depicted in Figure 6.2, where a digital buffer stage is driven by the V_{PDM} signal to operate the output receiver in Class-D.

3. Design Example

Based on the Log companding CMOS circuit strategy proposed in the previous section, a 100KHz Class-D output stage is implemented for typical hearing aid applications. In this case, a single slope (i.e. saw-tooth) oscillator has been chosen with design parameters I_{bias} = 1μA, I_{tun} = 100nA, C = 5pF and I_{th} = 4μA. All boxed devices in Figure 6.2 were sized at $10 \times \frac{40\mu\text{m}}{3\mu\text{m}}$ to ensure weak inversion operation for the intended range of currents. On the other hand, the rest of MOS transistors were designed with aspect ratios as large as allowed by the supply voltage to reduce technology mismatching and output offset.

The complete PDM circuit is depicted in Figure 6.4 and has been integrated in the same 1.2μm CMOS double-metal double-poly-Si process as the rest of examples in this work. Although not shown, output buffers were also designed specifically for this power stage. In any case, detailed buffering techniques can be found in [7].

Both BSIM3 simulated and experimental performances of the PDM circuit are reported in Table 6.1 under true battery supply operation (i.e. 1.1V to 1.5V). Finally, an experimental PDM signal and its corresponding acoustic output at the receiver are depicted in Figure 6.5.

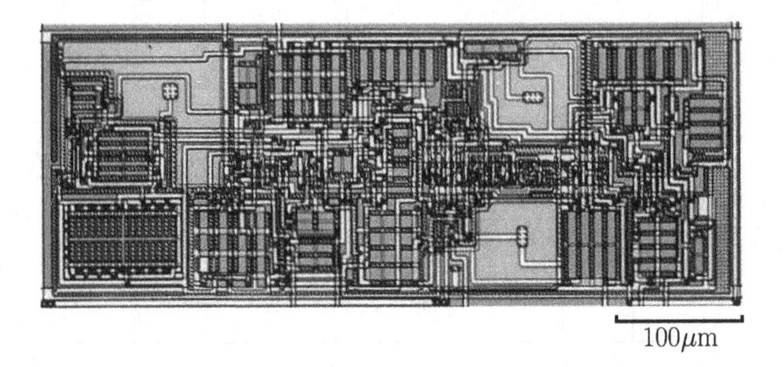

Figure 6.4. Microscope photograph of the PDM modulator.

Table 6.1. Typical results of the modulator at room temperature.

Parameter	Simulated	Experimental	Units
Min. Supply Voltage (including receiver)	1.1	1.1	V
Technology $(V_{TON}+\|V_{TOP}\|)_{max}$	1.3	not available	V
Input Full Scale (I_{max})	4	3.9	μA_{pp}
Total Harmonic Distortion @90%I_{max}	<2.8	<3	%
Dynamic Range @(100Hz-10KHz)	>70	>63	dB
Zero Input PDM frequency	102	104	KHz
PDM Factor	2.6	2.7	$\mu s/\mu A$
Power Consumption	45	not available	μW
Si Area	0.10	0.10	mm^2

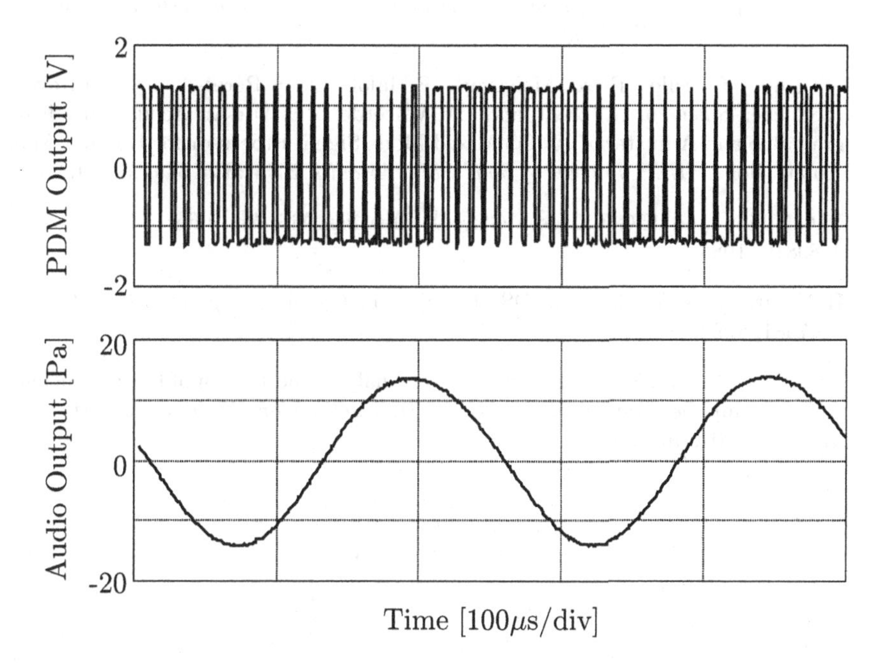

Figure 6.5. Experimental differential PDM voltage and acoustic audio signal at receiver for $I_{in}=2\mu A_{pp}$@4KHz.

References

[1] Standard Dictionary of Electrical and Electronics Terms. Standard 100-1988, IEEE, Nov 1988. http://www.ieee.org.

[2] K.Philips, J. van den Homberg, and C.Dijkmans. PowerDAC: A Single-Chip Audio DAC with a 70%-Efficient Power Stage in 0.5μm CMOS. In *Proceedings of the International Solid-State Circuits Conference*, pages 154–155. IEEE, 1999.

[3] M.T.Tan, J.S.Chang, Z.H.Cheng, and Y.C.Tong. A Novel Self-Error Correction Pulse Width Modulator for a Class D Amplifier for Hearing Instruments. In *Proceedings of the International Symposium on Circuits and Systems*, volume I, pages 261–264. IEEE, 1998.

[4] J.F.Duque-Carrillo, Piero Malcovati, F.Maloberti, R.Pérez-Aloe, A.H.Reyes, E.Sánchez-Sinencio, G.Torelli, and J.M.Valverde. VERDI: An Acoustically Programable and Adjustable CMOS Mixed-Mode Signal Processor for Hearing Aids Applications. *IEEE Journal of Solid State Circuits*, 31(5):634–645, May 1996.

[5] M.C.Killion and E.G.Village. Class D Hearing Aid Amplifier. U.S. Patent 4689819, 1985.

[6] H.A.Gurcan. Class D BiCMOS Hearing Aid Output Amplifier. U.S. Patent 5247581, 1993.

[7] J.S.Chang, M.Tan, Z.Cheng, and Y.Tong. Analysis and Design of Power Efficient Class D Amplifier Output Stages. *IEEE Transactions on Circuits and Systems-I*, 47(6):897–902, Jun 2000.

Chapter 7

DYNAMIC RANGE

Abstract The following sections present a general study about the accuracy of all CMOS circuit techniques proposed in previous chapters. In particular, basic design equations at the device level are derived from distortion and noise considerations. Also, different dynamic range versus signal-to-noise ratio strategies are discussed in the Log companding frame of this study.

1. CMOS Considerations

Many exhaustive analysis exist about both distortion and noise issues in Log companding circuits [1, 2, 3, 4, 5, 6, 7, 8, 9, 10, 11]. However, most of the synthesis techniques in this field are focused on bipolar technologies (i.e. BJT-based implementations). As a result, this section presents the particular considerations to be taken into account when using the MOSFET as the basic processor for Log companding.

The common scenario for the CMOS circuit techniques proposed in this work can be represented by Figure 7.1. In all cases, the linear input signal in the I-domain (I_{in}) is first biased (I_{biasi}) to ensure $I_{in} > 0$ and enable its Log mapping. Then, it is translated to the internal compressed V-domain by a compressor device (M1) using either a gate- (GD) or a source-driven (SD) companding function F. After a proper non-linear internal voltage processing, the expander (M2) ideally restores the original linearity and dynamic range at the output. Finally, a bias level (I_{biaso}) must be cancelled according to the overall signal gain.

Due to the wide variety of Log companding processing implemented in previous chapters, the following dynamic range study is mainly focused on the compression and expansion processes, common stages to all the proposed circuit techniques. Furthermore, a gain (G) will be considered

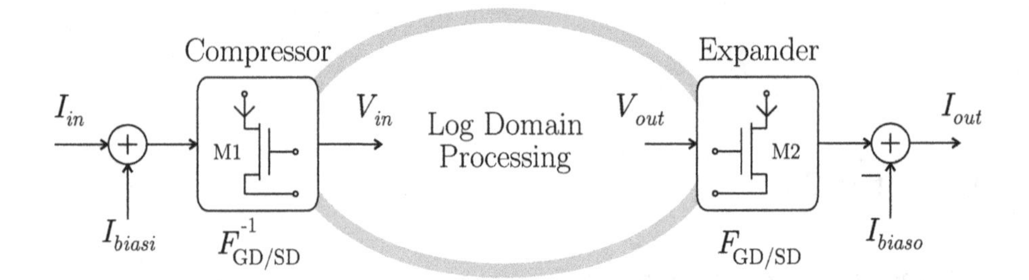

Figure 7.1. General CMOS Log companding processing scheme.

here in order to generalize the results for different input and output bias levels (i.e. $I_{biasi} \neq I_{biaso}$).

The key design parameters considered at this level are the channel dimensions of both the compressor and the expander devices ($W_{1,2}$ and $L_{1,2}$), which are supposed to be identical, and the bias levels (I_{biasi} and I_{biaso}). In fact, these variables can be respectively seen as the system specifications for Area and Power.

From a theoretical point of view, all the proposed strategies of this work ideally implement the desired companding processing as long as their main devices behave according to the exponential I/V characteristics of weak inversion detailed in Table 2.1. However, two physical effects set bounds to this model in actual circuits: moderate inversion and noise floor. The first non-ideal behaviour limits the maximum signal level based on distortion criteria, while the second effect defines the minimum signal amplitude which is not masked by internal random phenomena.

1.1 Moderate Inversion distortion

distortion in Figure 7.1 may occur even for signal values not reaching the saturation levels defined by $I_{biasi,o}$. This effect takes place when either M1 or M2 leave their weak inversion region, causing a degradation of the purely exponential I/V law towards the well-known quadratic characteristic in strong inversion. The undesired consequence in gain loss can be clearly seen in Figure 7.2, where V_{gain} represents the general tuning parameter in the V-domain as proposed in Chapter 3.

In order to evaluate the progressive distortion due to moderate inversion, the following unified expression of the drain current in saturation will be taken from Table 2.1:

$$I_D = I_S \ln^2 \left[1 + e^{\frac{V_{GB} - V_{TO}}{2nU_t}} e^{-\frac{V_{SB}}{2U_t}} \right] \tag{7.1}$$

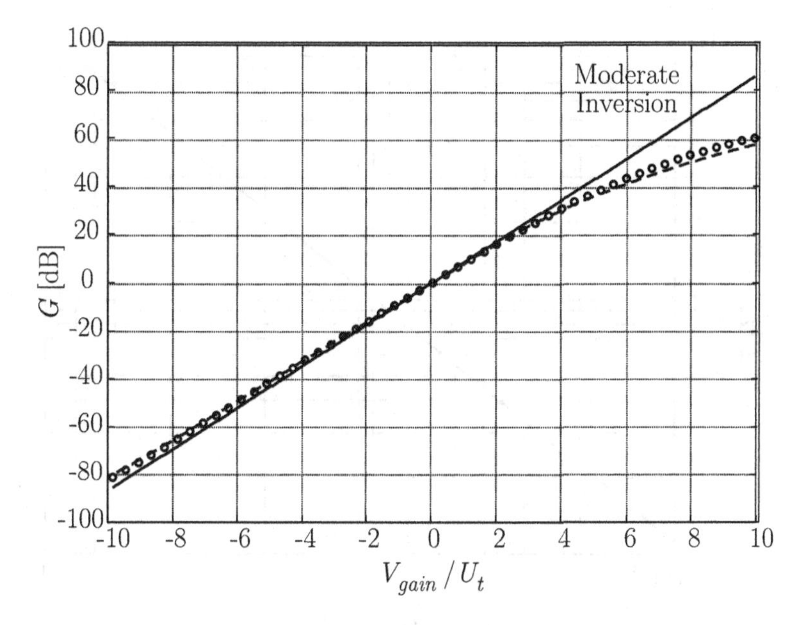

Figure 7.2. Experimental (dotted), simulated (dashed) and ideal (solid) DC signal gain for $IC_{in} = 10^{-2}$.

Based on this general equation, the input-output large-signal transfer function valid for all regions of operation and for both GD and SD compression laws is as follows:

$$IC_{out} = \ln^2 \left[1 + \sqrt{G}\left(e^{\sqrt{IC_{in}}} - 1\right)\right] \tag{7.2}$$

where the specific currents IC_{out} and IC_{in} are defined according to (4.57). Different situations may be considered at this point. For instance, the above equation returns $IC_{out} \equiv IC_{in}$ for $G = 1$, which is in fact the classic current mirror, perhaps the only distortion-free companding structure for all regions of MOS saturation. However, the worst case is usually related to large amplification factors ($G \gg 1$) with M1 still kept in deep subthreshold operation ($IC_{in} \ll 1$) but M2 already entering into its moderate inversion region ($IC_{out} \sim 1$). In these cases, the above expression (7.2) can be simplified to:

$$IC_{out} \simeq \ln^2 \left(1 + \sqrt{G IC_{in}}\right) \qquad IC_{in} \ll 1 \tag{7.3}$$

Now, the total harmonic distortion (THD) will be used to evaluate the amplitude-dependent non-linearity. In order to apply a harmonic analysis, the following polynomial interpolation is chosen:

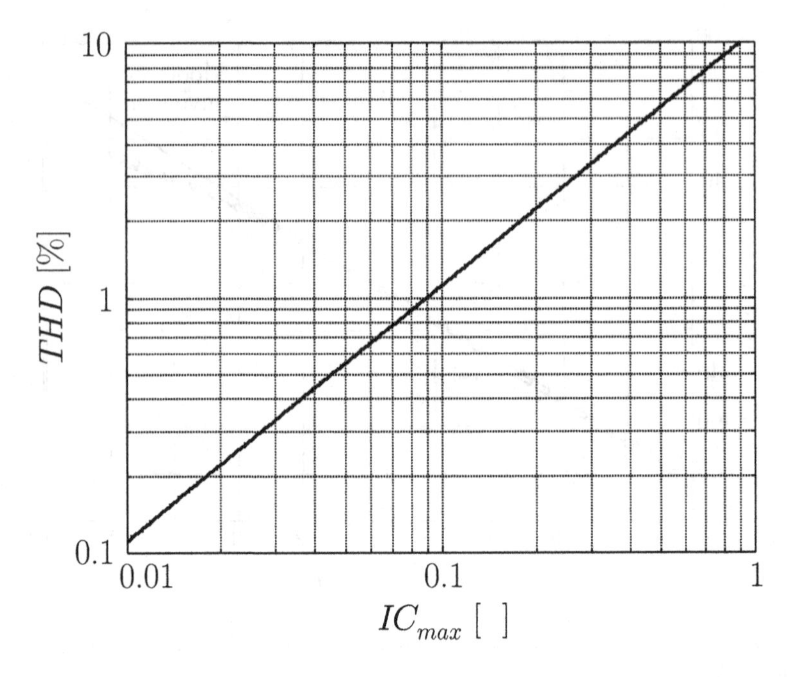

Figure 7.3. Analytical THD at 50% of full-scale due to moderate inversion.

$$IC_{out} \simeq \frac{3}{4}GIC_{in} - \frac{1}{2}\left(GIC_{in}\right)^2 + \frac{1}{5}\left(GIC_{in}\right)^3 \tag{7.4}$$

As a result, an approximate expression for THD at 50% of full-scale (I_{max}) is found to be:

$$THD[\%] \simeq 100\frac{IC_{max}}{9} = \frac{100}{9}\frac{I_{max}}{I_{Su}\left(\frac{W}{L}\right)_{1,2}} \tag{7.5}$$

where I_{Su} stands for the unitary specific current of the CMOS process following (2.4). This equation gives a simple rule to design the minimum aspect ratio $(W/L)_{1,2}$ of the Compressor and Expander devices for a given full-scale and distortion specification. Typically, wide aspect ratios (e.g. $(W/L)_{1,2} > 100$) are required for $THD < 1\%$ in order to keep I_{max} more than one decade below the specific current, as depicted in Figure 7.3. Also, the bias reference I_{max} should be obtained from a source proportional to I_S in order to keep the THD constant against temperature. In this sense, suitable Log companding CMOS techniques for such a synthesis are proposed in Chapter 5.

1.2 Noise Floor

In general, noise phenomena in MOS devices are composed of both thermal and flicker parts as explained in Chapter 2. In the context of Log companding, although processing is based on large-signal device models, a small-signal power spectral density (PSD) can be used under Class-A operation [6]. For weak inversion saturation, PSD model of the noise drain current is taken from (2.22):

$$\frac{\mathrm{d}{i_{DN}}^2}{\mathrm{d}f} = 4KT\frac{g_{ms}}{2} + \frac{K_{fk}}{WL}\frac{g_{mg}^2}{f} = 2qI_D + \frac{K_{fk}}{WL}\left(\frac{I_D}{nU_t}\right)^2\frac{1}{f} \tag{7.6}$$

where q and K_{fk} stand for the electron charge and the technological flicker constant, respectively. In order to quantify its effect in the general scheme of Figure 7.1, noise is integrated at the output linear I-domain:

$$\left(\frac{i_{Nout}}{I_{biaso}}\right)^2 = \frac{2q(1+G)}{I_{biaso}}\left(f_{up} - f_{low}\right) + \frac{2}{(nU_t)^2}\frac{K_{fk}}{(WL)_{1,2}}\ln\left(\frac{f_{up}}{f_{low}}\right) \tag{7.7}$$

where f_{up} and f_{low} define the noise bandwidth. Now, it is clear that the effect of the thermal component can be minimized by increasing the current biasing. On the other hand, the flicker part is directly proportional to the bias level. The latter behaviour is of special interest since signal full-scale (I_{max}) is also related to the same reference through $I_{max}[\mathrm{rms}] = I_{biaso}/\sqrt{2}$ due to Class-A operation. Hence, the CMOS Log companding system behaves as a signal processor with a resolution proportional to the signal room itself. As a general rule, the optimum solution is reached when thermal and flicker noise components are equal. Otherwise, in case of thermal or flicker dominance, the design is wasting area or power, respectively. Hence, the optimum biasing level for a given device area is obtained by:

$$I_{biaso} = \frac{q(1+G)(nU_t)^2}{K_{kf}}\frac{f_{up} - f_{low}}{\ln\left(f_{up}/f_{low}\right)}(WL)_{1,2} \tag{7.8}$$

Taking the maximum signal-to-noise ratio definition,

$$SNR[\mathrm{dB}] \doteq 20\log\frac{I_{max}[\mathrm{rms}]}{i_{Nout}} \tag{7.9}$$

the maximum value for the I_{biaso} case results in:

$$SNR_{\text{Class-A}}[\mathrm{dB}] = 10\log\frac{(WL)_{1,2}(nU_t)^2}{4K_{fk}\ln\left(\frac{f_{up}}{f_{low}}\right)} - 3 \tag{7.10}$$

2. Dynamic Range Versus Signal-to-Noise Ratio

Combining the two main equations from the previous section, specifications and design parameters of the CMOS Log companding scheme in Figure 7.1 are interrelated according to the following diagram:

$$\left.\begin{array}{ccc} SNR & \xrightarrow{(7.10)} & (WL)_{1,2} \\ & & \downarrow^{(7.8)} \\ & & I_{max} \equiv 2I_{biaso} \\ & & \downarrow^{(7.5)} \\ THD & \xrightarrow[(7.5)]{} & \left(\frac{W}{L}\right)_{1,2} \end{array}\right\} \quad W_{1,2} \text{ and } L_{1,2} \qquad (7.11)$$

In other words, the absolute dimensions for compressor and expander devices are derived by combining area requirements from noise, and aspect ratios from THD specifications. An experimental example of the above design procedure can be seen in Figure 7.4. In this case, compressor and expander are designed to meet $THD < 0.5\%$ at 50% full-scale, and $SNR > 75$dB from 100Hz to 10KHz when $G \equiv 1$. The resulting NMOS transistors for the same 1.2μm CMOS technology of Chapter 8 exhibit an aspect ratio of $(W/L)_{1,2} = 20 \times \frac{80\mu m}{3\mu m}$ and are operated at a full-scale of $I_{max} = 4\mu A_{pp}$. The specific current results in $I_{S1,2} = 75\mu A$, while the integrated output noise is about $i_{Nout} = 0.2nA_{rms}$. The particular analog layout techniques to maximize device matching between such a compressor and expander are based on common centroid structures, as explained in Section 5.

An interesting conclusion from the above design strategy is the fact that the Class-A operation automatically defines the maximum signal-to-noise ratio for a given device area. The equivalent dynamic range (DR) can be identified as:

$$DR_{\text{Class-A}} \simeq SNR_{\text{Class-A}} \qquad (7.12)$$

When requiring larger DR values, a different class operation may be introduced. This problem has already been faced in previous bipolar implementations by choosing a Class-AB scheme [12, 13, 14, 15, 16, 17, 18]. In the MOSFET case, the original SNR expression in (7.10) would be rewritten as:

$$SNR_{\text{Class-AB}}[\text{dB}] = 10 \log \frac{(WL)_{1,2}(nU_t)^2}{4K_{fk} \ln\left(\frac{f_{up}}{f_{low}}\right)} + 20 \log \frac{MI_{bias}/\sqrt{2}}{I_{bias}} \qquad (7.13)$$

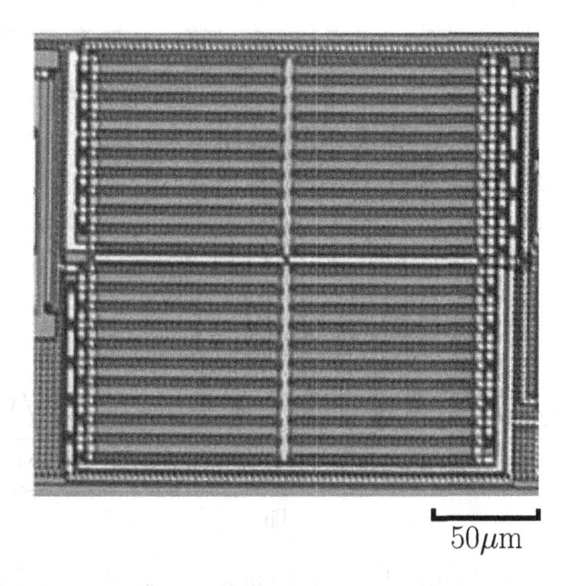

Figure 7.4. Microscope photography of a $20\times\frac{80\mu m}{3\mu m}$ NMOS compressor and expander pair example with a common centroid layout symmetry for a 1.2μm CMOS technology.

resulting in:

$$DR_{\text{Class-AB}}[\text{dB}] \simeq SNR_{\text{Class-A}} + 20\log M \tag{7.14}$$

where M stands for the ratio between the maximum input signal and the quiescent bias level of M1 and M2. However, from the companding point of view, these solutions exhibit $I_{max} > I_{bias}$, thus requiring us to map the negative left half-plane of signals (i.e. $I < 0$) in the Log domain. In order to overcome such a problem, Class-AB bipolar implementations make use of input splitters [12, 14, 15, 16, 18] or hyperbolic F functions [13, 14, 17]. However, drawbacks arise then due to: splitting non-linearity, Si area overhead and more complex compressors and expanders.

The operation class proposed here aims to take advantage of Class-A linearity and compact area while allowing larger DR by using a non-DC biasing level. Unlike previous static modes, the basic idea is to use dynamic bias adapted in time to signal demands: $I_{biasi,o}$ is increased or decreased according to signal envelope variations as illustrated in Figure 7.5. This adaptive Class-A mode is also referred to in some environments as Class-H [19].

Although such biasing control exhibits the same maximum signal-to-noise ratio as the Class-A from (7.10), this value is kept constant as long

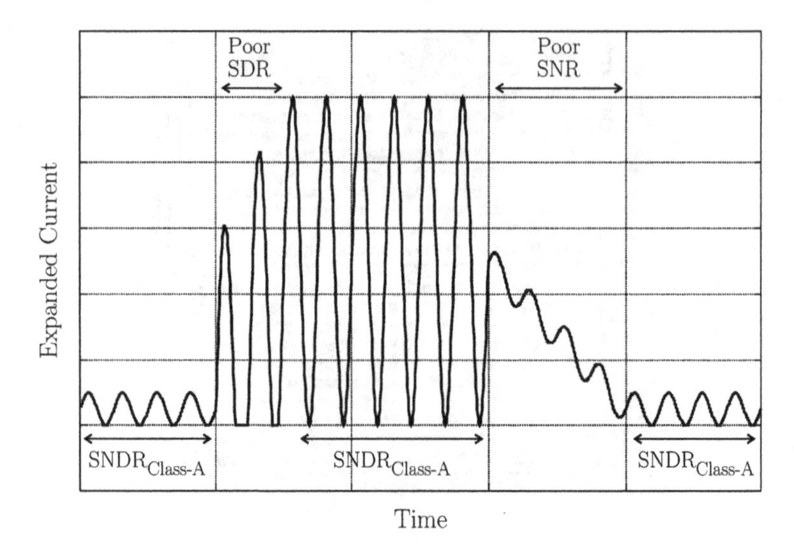

Figure 7.5. Example of a transient waveform in a Class-H operation. Drawing not in scale.

as $I_{biasi,o}$ are adapted to the steady-state levels of $I_{in,out}$, respectively. Then, DR is no longer limited by the available SNR but spanned to:

$$\begin{aligned} DR_{\text{Class-H}}[\text{dB}] &= 10 \log \frac{(WL)_{1,2}(nU_t)^2}{4K_{fk} \ln\left(\frac{f_{up}}{f_{low}}\right)} + 20 \log \frac{I_{biasmax}/\sqrt{2}}{I_{biasmin}} \\ &= SNR_{\text{Class-A}} + 20 \log M \end{aligned} \tag{7.15}$$

where M stands for the maximum biasing scaling. In other words, the novel biasing proposal uses a fixed signal resolution within the adapted range. The steady-state behavior of Class-H can be easily seen in Figure 7.6. Flat SNR corresponds to the adapted I_{bias} range, which clearly increases the final DR in comparison to Class-A and also eliminates the unnecessary extra SNR supplied by Class-AB.

In practice, the available M scaling is limited by thermal noise, which follows a $\sqrt{I_{bias}}$ dependence as marked in the same figure. In fact, the resulting corner is the optimum design point according to the design flow (7.11) for the static Class-A approach.

Unfortunately, Class-H can only be directly applied to the amplification techniques of Chapter 3, since very fast internal time constants are assumed for these blocks. On the other hand, some extra circuitry would be required to update the internal state-space (SS) variables of the system for the general filtering structures of Chapter 4, which usu-

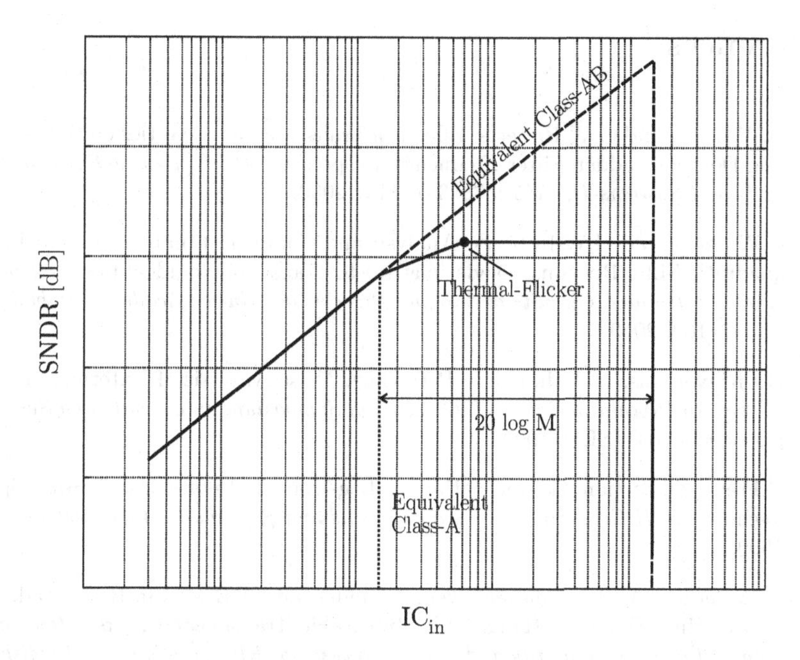

Figure 7.6. Example of signal-to-noise-and-distortion ratio versus input inversion coefficient for equivalent Class-A (dotted), Class-AB (dashed) and Class-H (solid) operation.

ally have time constants similar to the control of Class-H operation, as pointed out in [20, 21, 22].

Furthermore, the proposed Class-H technique is only suitable when processing information to be finally analyzed by non-instantaneous adaptive receivers, that is, in practice, for human perception applications (e.g. voice, audio and video interfaces). Also, time constants of the $I_{biasi,o}$ dynamic control should be tuned to the adaptive speed and instantaneous sensitivity of the particular receiver (e.g. from 10ms to 1s and about 45dB for human hearing [23, 24, 25, 26]).

As a result, the Class-H alternative to Class-A operation requires further research, which escapes the scope of this study.

References

[1] V.W.Leung and G.W.Roberts. Effects of Transistor Nonidealities on High-Order Log-Domain Ladder Filter Frequency Responses. *IEEE Transactions on Circuits and Systems-II*, 47(5):373–387, May 2000.

[2] V.W.Leung and G.W.Roberts. Analysis and Compensation of Log-Domain Biquadratic Filter Response Deviations due to Transistor Nonidealities. *Journal of Analog Integrated Circuits and Signal Processing, Kluwer Academic Publishers*, 22:147–162, 2000.

[3] G.Efthivoulidis, L.Tóth, and Y.P.Tsividis. Noise Analysis of Externally Linear Filters. In *Proceedings of the International Symposium on Circuits and Systems*, pages 960–963. IEEE, 1999.

[4] D.Frey. Distortion Compensation in Log-Domain Filters Using State-Space Techniques. *IEEE Transactions on Circuits and Systems-II*, 46(7):860–869, Jul 1999.

[5] W.A.Serdijn, M.H.L.Kouwenhoven, J.Mulder, and A.H.M. van Roermund. Design of High Dynamic Range Fully Integrable Translinear Filters. *Journal of Analog Integrated Circuits and Signal Processing, Kluwer Academic Publishers*, 19(3):223–239, Jun 1999.

[6] J.Mulder, M.H.L.Kouwenhoven, W.A.Serdijn, A.C. van der Woerd, and A.H.M. van Roermund. Nonlinear Analysis of Noise in Static and Dynamic Translinear Circuits. *IEEE Transactions on Circuits and Systems-II*, 46(3):266–278, Mar 1999.

[7] L.Tóth, Y.P.Tsividis, and N.Krishnapura. On the Analysis of Noise and Interface in Instantaneously Companding Signal Processors. *IEEE Transactions on Circuits and Systems-II*, 45(9):1242–1249, Sep 1998.

[8] J.Mulder, M.H.L.Kouwenhoven, W.A.Serdijn, A.C. van der Woerd, and A.H.M. van Roermund. Noise Considerations for Translinear Filters. *IEEE Transactions on Circuits and Systems-II*, 45(9):1199–1204, Sep 1998.

[9] M.Punzenberger and C.C.Enz. Noise in Instantaneous Companding Filters. In *Proceedings of the International Symposium on Circuits and Systems*, volume 1, pages 337–340. IEEE, Jun 1997.

[10] J.Mulder, M.H.L.Kouwenhoven, and A.H.M. van Roermund. Signal × Noise Intermodulation in Translinear Filters. *IEE Electronics Letters*, 33(14):1205–1207, Jul 1997.

[11] V.W.Leung, M.El-Gamal, and G.W.Roberts. Effects of Transistor Nonidealities on Log-Domain Filters. In *Proceedings of the International Symposium on Circuits and Systems*, volume 1, pages 477–480. IEEE, Jun 1997.

[12] A.T.Tola and D.R.Frey. A Study of Different Class AB Log Domain First Order Filters. *Journal of Analog Integrated Circuits and Signal Processing, Kluwer Academic Publishers*, 22:163–176, 2000.

[13] J.Mahattanakul and C.Toumazou. Modular Log-Domain Filters Based upon Linear Gm-C Filter Synthesis. *IEEE Transactions on Circuits and Systems-I*, 46(12):1421–1430, Dec 1999.

[14] D.R.Frey and A.T.Tola. A State-Space Formulation for Externally Linear Class AB Dynamical Circuits. *IEEE Transactions on Circuits and Systems-II*, 46(3):306–314, Mar 1999.

[15] P.J.Poort, W.A.Serdijn, J.Mulder, A.C. van der Woerd, and A.H.M. van Roermund. A 1-V Class-AB Translinear Integrator for Filter Applications. *Journal of Analog Integrated Circuits and Signal Processing, Kluwer Academic Publishers*, 21:79–90, 1999.

[16] M.Punzenberger and C.C.Enz. A 1.2-V Low-Power BiCMOS Class AB Log-Domain Filter. *IEEE Journal of Solid State Circuits*, 32(12):1968–1978, Dec 1997.

[17] D.R.Frey. Exponential State Space Filters: A Generic Current Mode Design Strategy. *IEEE Transactions on Circuits and Systems-I*, 43(1):34–42, Jan 1996.

[18] D.R.Frey. Current Mode Class AB Second Order Filter. *IEE Electronics Letters*, 30(3):205–206, Feb 1994.

[19] Gennum Corp., 970 Fraser Drive, Burlington, Ontario L7L 5P5. *GS3013 Application Note: Current Mode Class-H Amplifier*, Sep 1994. http://www.gennum.com.

[20] N.Krishnapura, Y.Tsividis, and D.R.Frey. Simplified Technique for Syllabic Companding in Log-Domain Filters. *IEE Electronics Letters*, 36(15):1257–1259, Jul 2000.

[21] C.C.Enz and E.A.Vittoz. CMOS Low-Power Analog Design. In *Emerging technologies: Designing Low Power Digitals Systems*, pages 79–133. IEEE, 1996.

[22] E.M.Blumenkrantz. The Analog Floating Point Technique. In *Symposium on Low Power Electronics*, pages 72–73. IEEE, 1995.

[23] F.Baumgarte. A Physiological Ear Model for Auditory Masking Applicable to Perceptual Coding. In *103rd AES Conference*, volume Preprint 4511, New York, Sep 1997. Audio Engineering Society.

[24] T.Dau. *Modeling Auditory Processing of Amplitude Modulation*. PhD thesis, Oldenburg University, Feb 1996. http://www.bis.uni-oldenburg.de/bisverlag/daumod96/daumod96.html.

[25] M.Slaney and R.F.Lyon. *On the Importance of Time - A Temporal Representation of Sound*, chapter 5, pages 95–116. John Wiley & Sons Ltd., 1993. http://www.pcmp.caltech.edu/anaprose/dick/.

[26] C.Mead. *Analog VLSI and Neural Systems*. Addison-Wesley Publishing Company, 1989.

Chapter 8

INDUSTRIAL APPLICATION: HEARING AIDS

Abstract This chapter presents a complete example on the application of all the new circuit techniques proposed in this work. The novel CMOS subthreshold topologies for Log amplification, filtering, generation and modulation are used here to implement a hearing-aid-on-chip for an industrial customer. Overall specifications for the application specific integrated circuit (ASIC) are described as well as the design flow for its full-custom implementation. Advantages of the new circuit approach are discussed and compared to other similar products.

1. History and Market

Hearing aids (HAs) are perhaps the clearest product example where users want to profit from the miniaturization and power-saving offered by the latest technologies in microelectronics.

Historically, HAs have always made use of the newest technological inventions and the state-of-the-art design techniques to achieve the comfort and esthetic characteristics demanded by final users. This fact can be clearly seen in the chronological evolution of HA technology during the last century [1, 2] depicted in Figure 8.1. First electrical (also called carbon) HAs were based on the telephone principle invented by A.Graham Bell (1876). Those systems directly coupled the input carbon microphone to the output earphone, operating with a single battery from 3V to 6V. Hence, amplification was mostly limited to the sizing ratio between both transducers. Not much later than the triode vacuum tube invention by L.de Forest (1906), the first intermediate amplifying stages, based on a single device, were applied to HAs. Two batteries were necessary in those products to bias and heat the vacuum tube, respectively. As soon as the first application-specific tubes were available, HAs became more compact and wearable. However, the invention of the

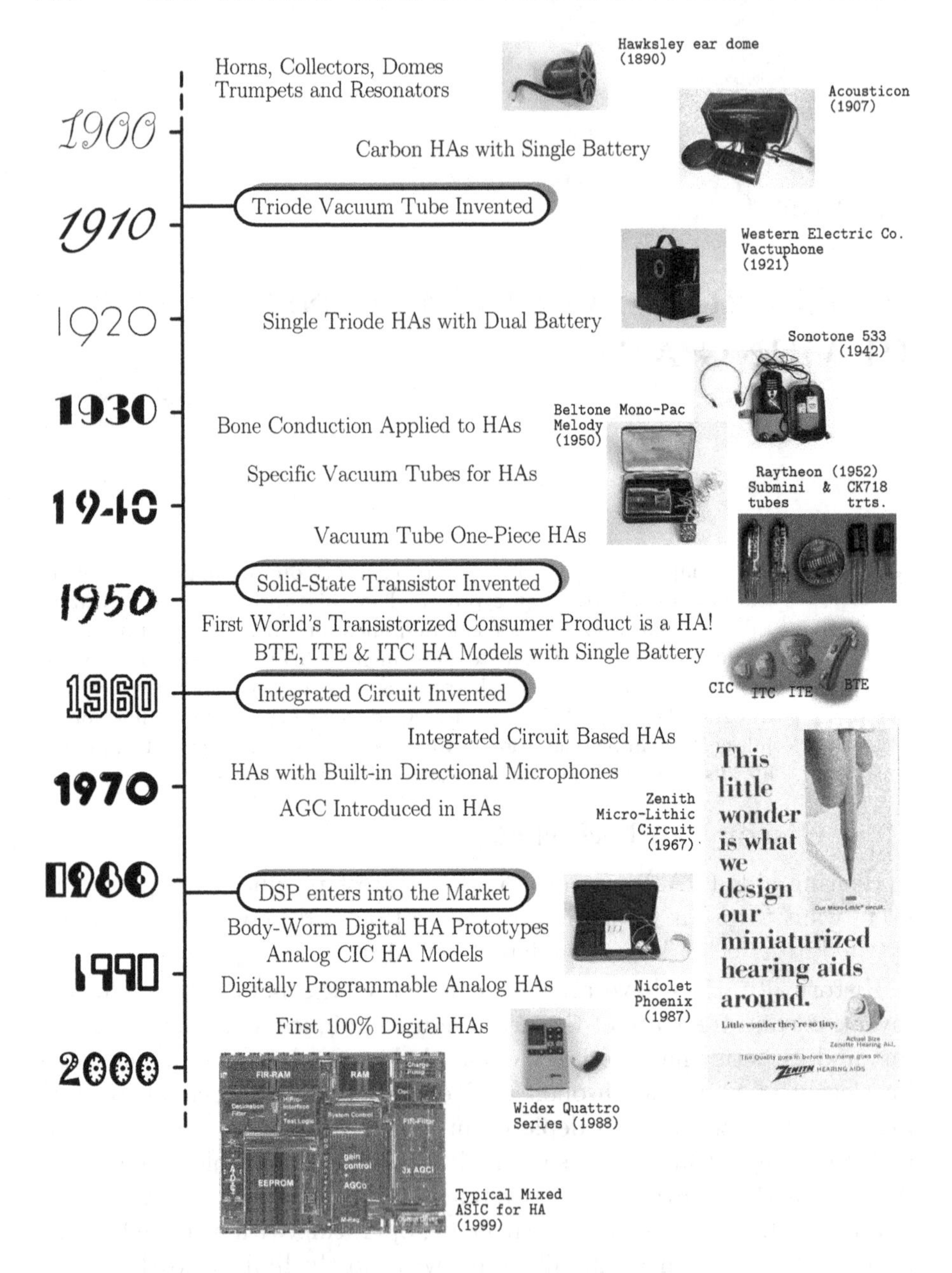

Figure 8.1. Chronology of HA products and technology evolution [1, 2].

solid-ttate transistor by J.Bardeen, W.Brattain and W.Shockley (1948) really caused a quantum leap in terms of power efficiency and reliability. In fact, the world's first transistorized consumer product was a HA.

The consequent size reduction led to the first portable behind-the-ear (BTE), in-the-ear (ITE), in-the-channel (ITC) models with a single small battery. Also, when the monolithic integrated circuit was invented by J.S.Killby (1958), HA manufacturers quickly applied this new technology to their products, allowing in that case an increase in the complexity of processing circuitry, like the first AGC stages. The last decades of the century have been dominated by the explosion of CMOS technology, which has been successfully exploited by HAs in terms of digitally programmable complex analog processors. This type of control increases the possibilities of fitting parameters to particular user losses. During the last five years, the first 100% digital HAs have also been launched to the market. These systems require A/D and D/A conversion stages, while signal processing is performed in the digital domain (DSP). dynamic range corrections are similar to the analog programmable products but audio processing is split here into different bands (typically from 2 to 10). Generally, technological options for the analog parts range from purely bipolar to CMOS implementations, all using a full-custom system-on-chip approach. Also, specific micro-packaging technologies are required, which combine flip-chip and flexible boards for Surface Mounted Devices (SMD).

Nowadays, both analog programmable and fully digital products share the HA market [3, 4, 5]. Advantages of the former are lower cost and usually lower power consumption, while the latter can incorporate more complex audiologic algorithms and exhibit a shorter time to market. In any case, critical design constraints in both cases come from battery technology, which imposes very low-voltage operation (down to 1.1V) and also low-power (below 0.5mA) in order to increase battery life (typically one week).

2. Previous CMOS Analog Systems

In the case of HAs based on CMOS technologies, circuit techniques can range from continuous-time active-RC to switched capacitors (SC) approaches depending on the particular technology. In any case, all the CMOS programmable analog designs reported in literature [6, 7, 8, 9, 10, 11, 12] avoid the very low-voltage constraints of HAs by making use of supply multipliers based on charge pumps as pointed out in Chapter 1. Typically, doublers are included to boost the 1.25V potential supplied by the battery, so that actual internal analog operation is at about 2.5V. Even most of the analog frontends (i.e. A/D converters) of full digital processing HAs make use of such circuit techniques [13]. However, supply multipliers tend to increase the die area as well as the number of discrete components required around the system-on-chip. Furthermore, power efficiency may be reduced in a product in which battery life is a critical specification. In this sense, the next section presents a novel HA product based on all the new very low-voltage CMOS circuit techniques proposed in this work, which avoids the necessity of such supply multipliers. Thanks to the true low-voltage internal operation, the overall current consumption of the system-on-chip can be strongly minimized as well.

3. A True 1V CMOS Log-Domain Analog Hearing-Aid-on-Chip

This section describes in brief the development of NEXO©, a system-on-chip processing core for a full family of analog HAs, as a complete application example of all the Log companding circuit techniques proposed along Chapters 3 to 6. This project is a joint venture between Microson S.A., as the industrial customer, and Centro Nacional de Microelectrónica (CNM), as the design center.

3.1 System-on-Chip Specifications

The customer request can be summarized as an analog device for processing audio signals which can operate at supply voltages as low as 1.1V together with a very low-current consumption. Also, electrical performance and programmable capabilities of that ASIC must fit the complete range of HA products (i.e. BTE, ITE, ITC and completely-in-the-channel CIC). Finally, CMOS technology are preferred in order to ensure manufacturing alternatives and to minimize costs. From the processing point of view, the system should include the following blocks, as depicted in Figure 8.2:

Transducer Front-end. This first stage is devoted to adapting the incoming signal levels from different input sources (i.e. electret microphone, telecoil and direct audio connector) and mix them together according to programmable weights. Also, a voltage regulator is required for the voltage supply of the different transducer drivers (e.g. JFET-resistor in the case of the electret microphone).

Curvilinear and Adaptive AGC-I to modify signal dynamic range according to user loss. The input-output transfer function is defined through three programmable parameters: Linear gain, compression ratio and threshold knee point. A combined fast and slow envelope detection is chosen here to synthesize attack and release times adapted to burst duration.

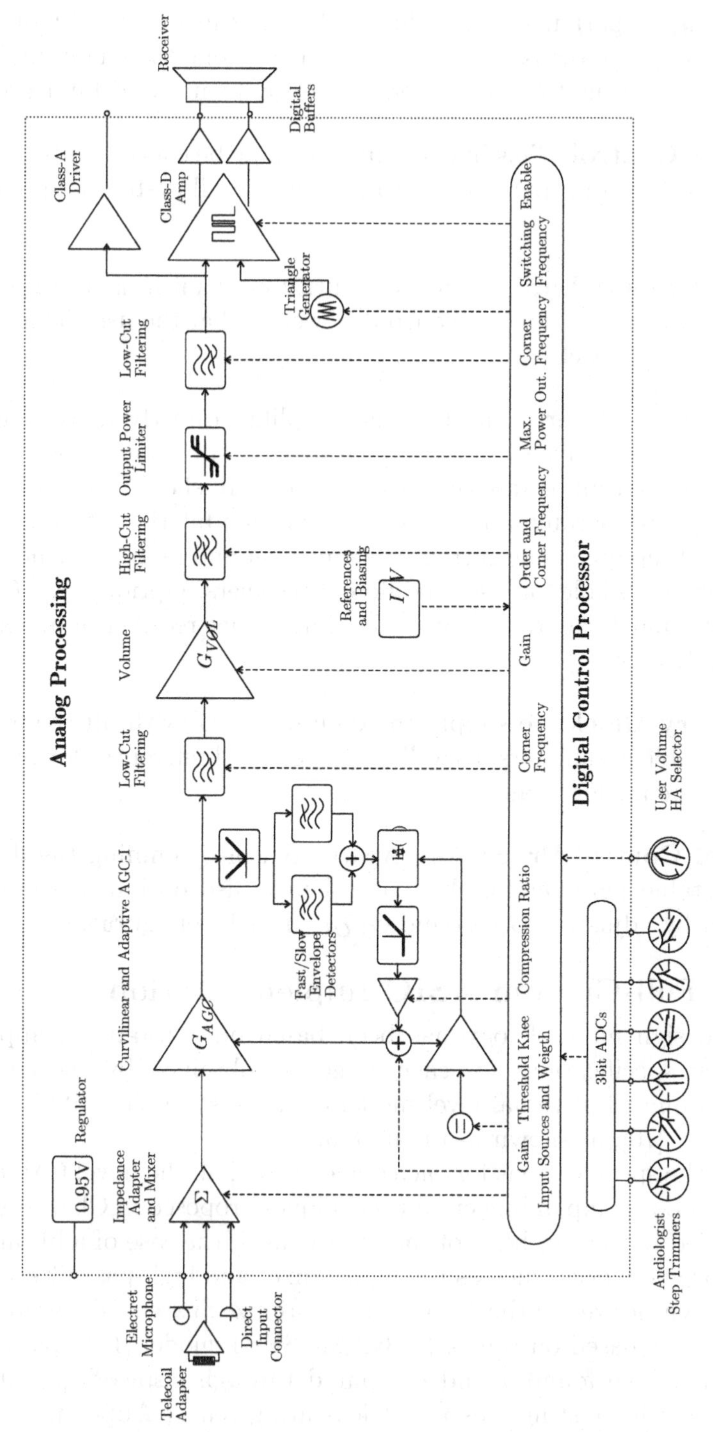

Figure 8.2. General block description of the HA.

Filtering. Apart from the inherent filtering in the AGC loop, some high-pass and low-pass filters are inserted along the chain, making it possible to tune the parameters corner frequency and filter order.

Volume Control. This linear gain stage may be controlled by the user and the program parameters to set the overall system amplification factor.

Output Power Limiter. Some security mechanism is also necessary in this system to limit the output power so that the user pain threshold is never reached.

Class-D amplifier. The backend amplifier directly drives the low-impedance receivers and compensates the differences between input and output transducer sensitivities. It should be based on a frequency-programmable pulse duration modulation strategy in order to keep quiescent current consumption and high-frequency losses as low as possible for a set of different receivers. Optionally, a Class-A output may be used to drive a high impedance receiver or an external Class-D driver.

Auxiliary Blocks to supply the digital control with all the external adjustable parameters, as well as the required references for programming the analog chain.

Digital Control Processor. Apart from programming the different adjustable parameters, this control unit also optimizes the overall SNR by adjusting G_{AGC} and G_{VOL} in each configuration.

3.2 Full-Custom ASIC Implementation

The design methodology has been based on a top-down approach. Electrical specifications for each stage are obtained by the customer himself from a functional-level model of the system simulated through Simulink© [10], as shown in Figure 8.3.

Once the specifications for each stage are set, all the novel CMOS subthreshold Log companding circuit techniques proposed in Chapters 3 to 6 are applied to the synthesis of each block, as in the case of a library cell. Some specific blocks, such as the input mixer and high-pass filtering, are directly synthesized in the I-domain to optimize Si area. Transistor-level simulation is based on the BSIM3v2 MOSFET model [1, 2] parameters supplied by the foundry and computed through Hspice© [3]. In this sense, some interesting tips and tricks are given in Appendix A. The previous functional-level model is also useful during this design step

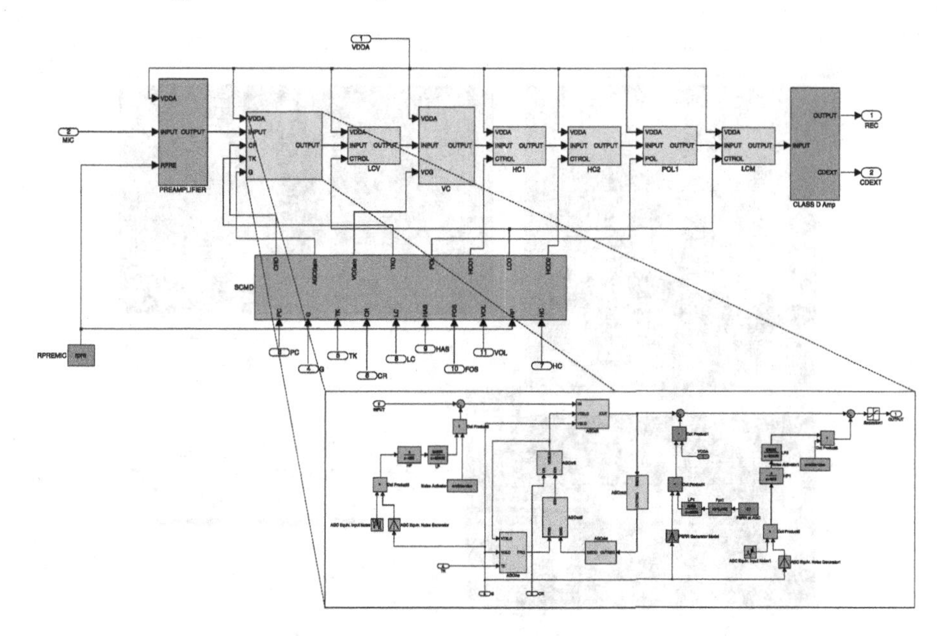

Figure 8.3. Functional-level model of the hearing-aid-on-a-chip.

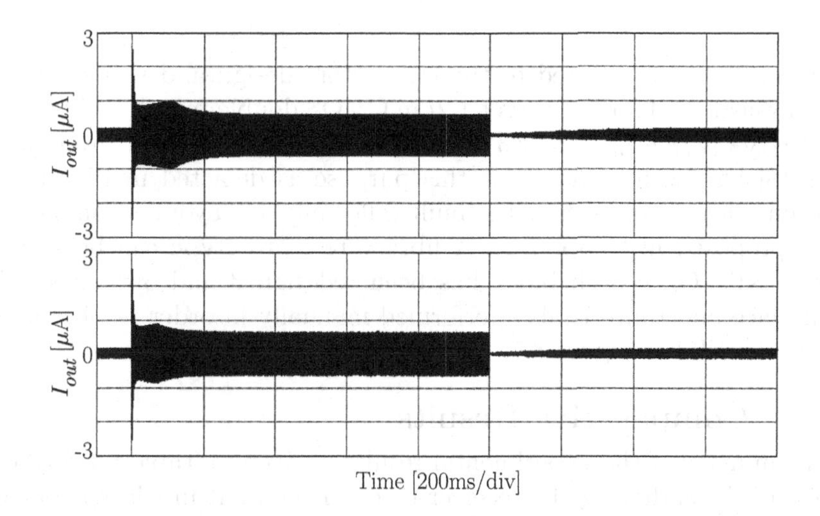

Figure 8.4. Functional (upper) and transistor (lower) level AGC simulation samples (CPU-time ratio was 15'/1500').

in order to save CPU-time when quantizing the effect of circuit non-idealities at the system level, such as noise and $PSRR$. An example of this design approach is illustrated in Figure 8.4, where functional and transistor-level (i.e. 5K devices) numerical outputs are compared.

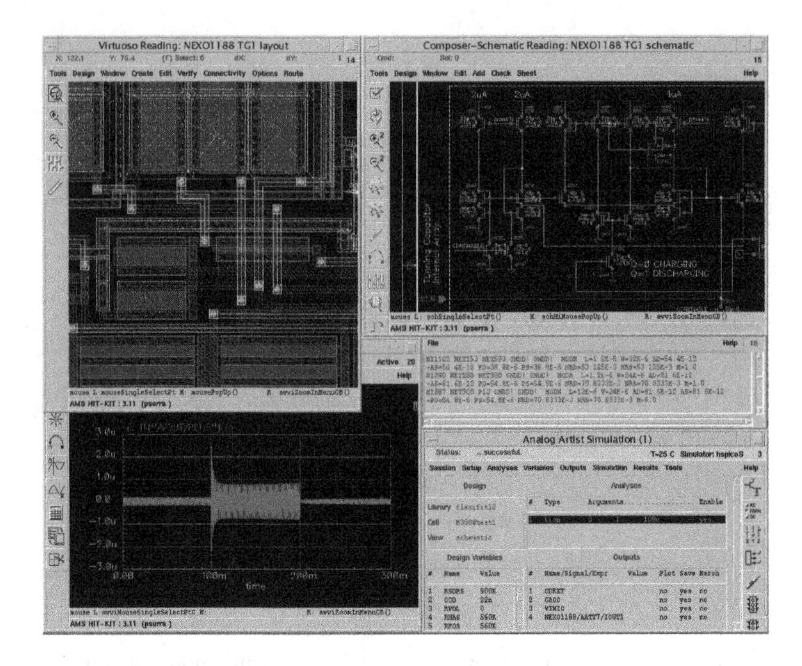

Figure 8.5. Full-custom design framework for the ASIC.

A third step is devoted to the full-custom design and verification at the physical-level for the target 1.2μm CMOS double-metal double-poly-Si $(V_{TON}+|V_{TOP}|)_{max}$ = 1.3V CMOS technology. Design Framework II© [18] has been selected for this purpose as depicted in Figure 8.5. The final layout of each cell is built following the layout recommendations proposed in Chapter 2. A library test prototype can be seen in Figure 8.6. Once each block has been redesigned and optimized, the final place and route is also performed manually in order to obtain the compact ASIC of Figure 8.7.

3.3 Comparative Results

A summary of the experimental results is given in Table 8.1 for both configurations, driving the receiver through the built-in Class-D output amplifier, or using an external Class-D by means of the additional Class-A output included in the ASIC. A qualitative comparison to other similar HA products, in terms of processing complexity and programmability, is also included in the same table. The overall gain and its fine digital-programmability in 3dB steps can be seen in Figure 8.8. The system also allows coarse digital control of such parameter in a range of 60dB. Figures 8.9 and 8.10 show the tuning capabilities of the low- and high-pass filter banks through different codes, respectively. Although not shown,

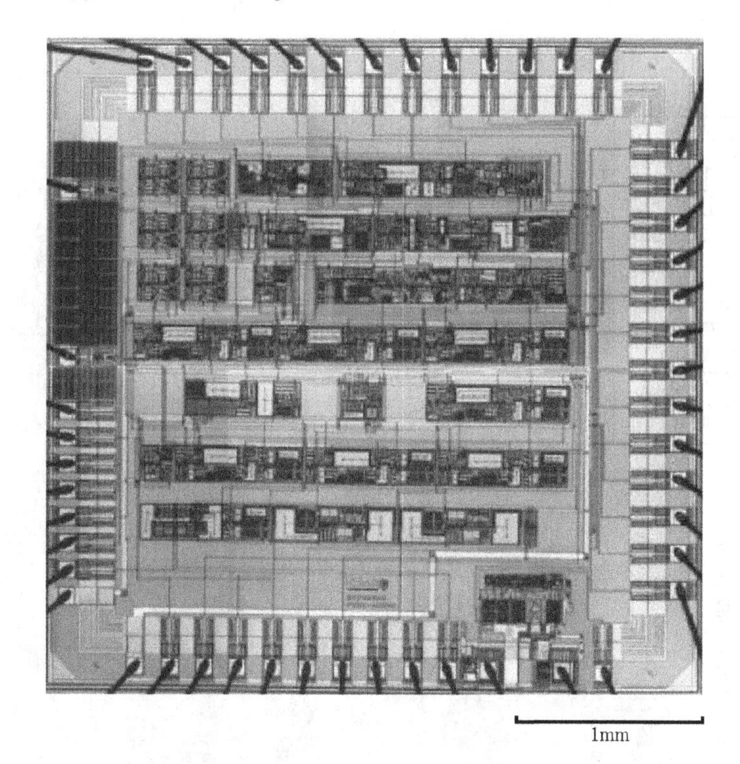

Figure 8.6. Microscope photography of a preliminary library prototype.

such control also includes order selection for the filtering. The flexibility of the adaptive AGC stage is illustrated in Figure 8.11, where the compression ratio, the threshold knee point and the open-loop gain are independently changed according to the digital control. Furthermore, the dual behaviour of the same AGC against perturbation-duration to keep speech intelligibility can be clearly seen in Figure 8.12. Finally, for the Class-D output stage, switching frequency can be programmed from 80KHz to 150KHz, while power buffers can drive up to $20mA_{peak}$ across the bridge. More detailed information about this design may be requested from the customer [19].

The reported results of this HA-on-chip in terms of high-gain, low-power, low-area, low-distortion, high-flexibility and low-cost performances meet the targeted specifications. Hence, using this single IC, all the HA products (i.e. behind-the-ear, in-the-ear, in-the-channel and completely-in-the-channel models) can be built changing only programming and some external elements like transducers. In fact, the novel analog CMOS circuit techniques for Log domain processing have allowed us to introduce significant innovations for the HA product. To the authors´ knowl-

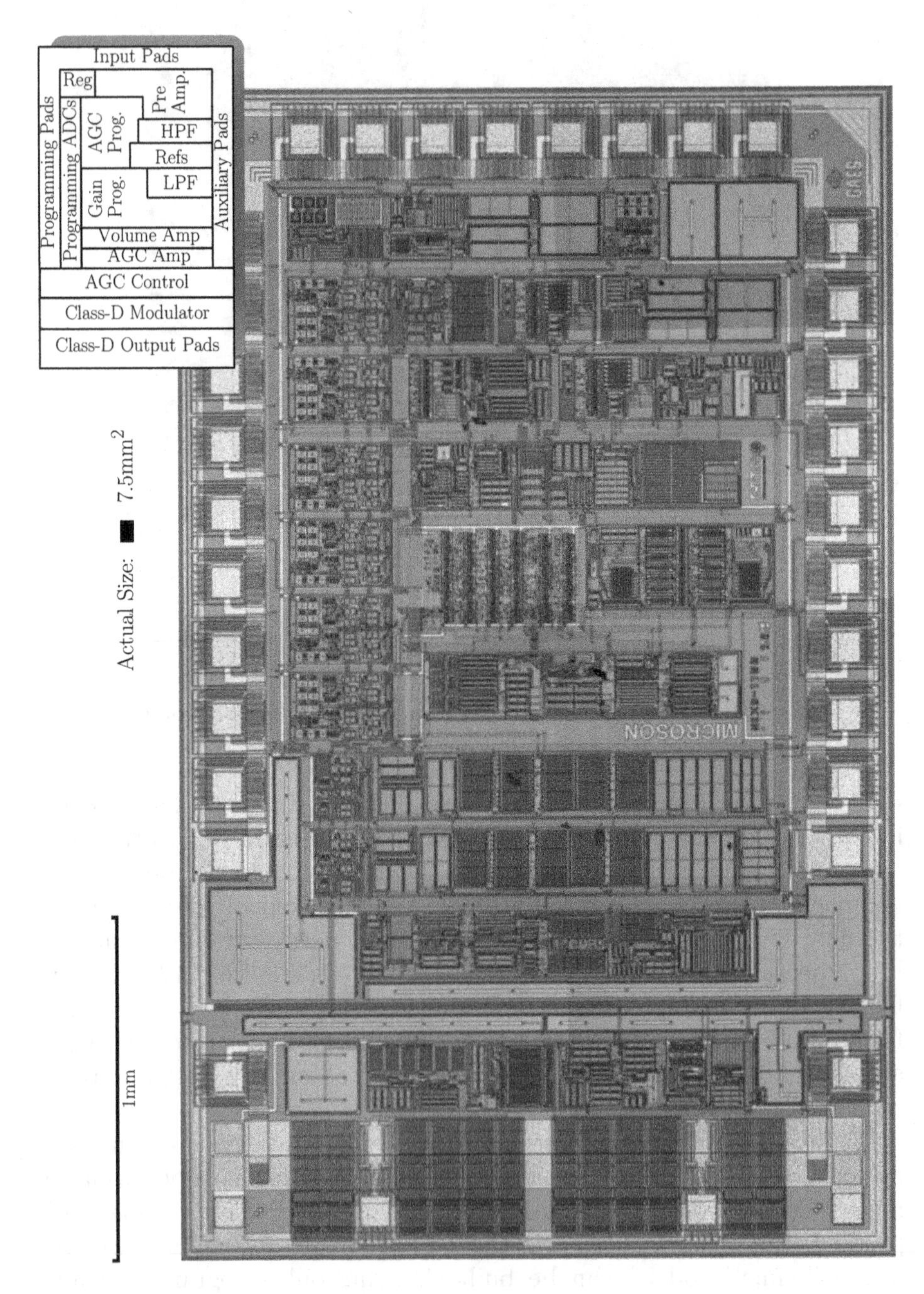

Figure 8.7. Microscope photography of the industrial hearing-aid-on-chip product.

Table 8.1. Preliminary NEXO© results and comparison to similar HA products, where V_{DDmin}, I_{DDQ}, R_{load}, G_{max}, V_{nieq} and THD stand for the minimum supply voltage, quiescent current consumption, load resistance, maximum electrical gain, equivalent input noise and total harmonic distortion, respectively. A $-40\text{dB}_{\text{V/Pa}}$ microphone sensitivity has been supposed. All HAs exhibit similar bandwidth (typically from 100Hz to 8KHz).

Parameter	This design	[20]	[21]	[22]	Units
HA models	BTE,ITE ITC,CIC	CIC	BTE	BTE,ITE ITC,CIC	-
Technology	CMOS	Bipolar	CMOS	CMOS	-
V_{DDmin}	1.0	1.1	1.1(×2)	1.1(×2)	V
I_{DDQ} $R_{load} \gg 100\Omega$	0.3	0.4	0.8	1.1	mA
$R_{load} \ll 100\Omega$	<0.5	no Class-D	<1.5	<1.4	
G_{max}	70	48	70	58	dB
V_{nieq} @(100Hz-10KHz)	6	3	5	7	μV_{rms}
THD w Class-D	<1	no Class-D	1	5	%
w/o Class-D	0.1	0.6	no Class-A	no Class-A	

edge, this is the first analog CMOS circuit, for either programmable HAs or for an A/D front-end in digital HAs, to truly operate at battery voltage (i.e. down to 1.1V), without the need of a charge-pump. In terms of energy, the final system exhibits one of the lowest current consumption in the HA market, as can be seen from Table 8.1. Another interesting feature is that the resulting implementation fits the electrical specifications from BTE (i.e. high-gain and low-impedance output) to CIC (i.e. low-noise and small-area) products on a single chip. Regarding technology constraints, the used circuit techniques do not require expensive sub-micron processes and can be implemented in a wide variety of CMOS technologies. Finally, the developed circuit topologies may be reused in the design of A/D frontends for digital HAs. In particular, the proposed filtering techniques seem very suitable for the synthesis of very low-voltage oversampling $\Sigma\Delta$ A/D converters as pointed out in Chapter 9.

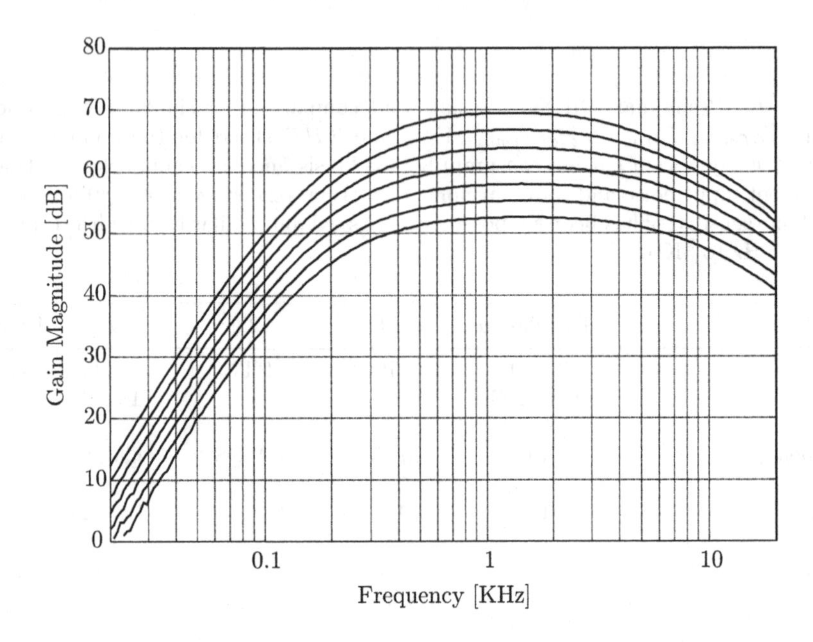

Figure 8.8. Experimental frequency transfer functions versus system gain.

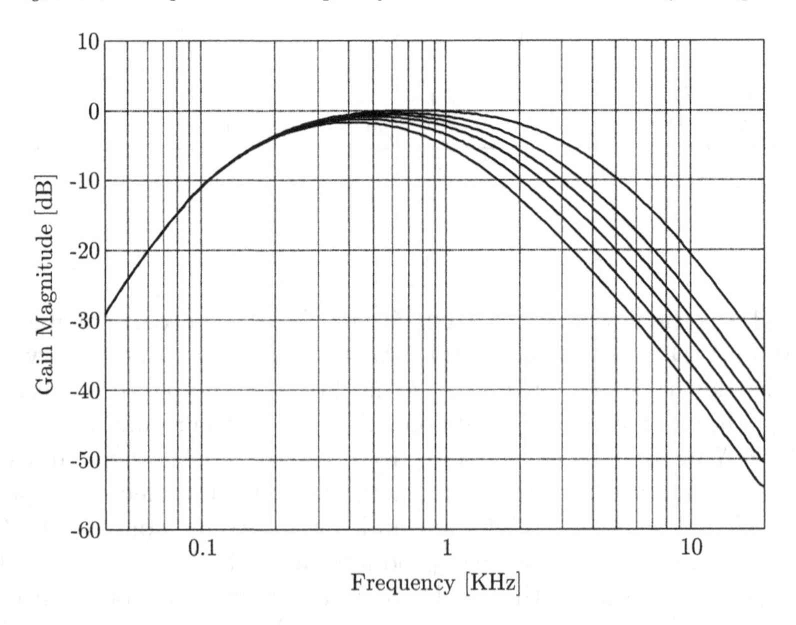

Figure 8.9. Experimental normalized transfer functions versus low-pass tuning.

4. Yield Issues

The industrial applicability presented in this chapter of all the new CMOS Log circuit techniques would not have much significance without

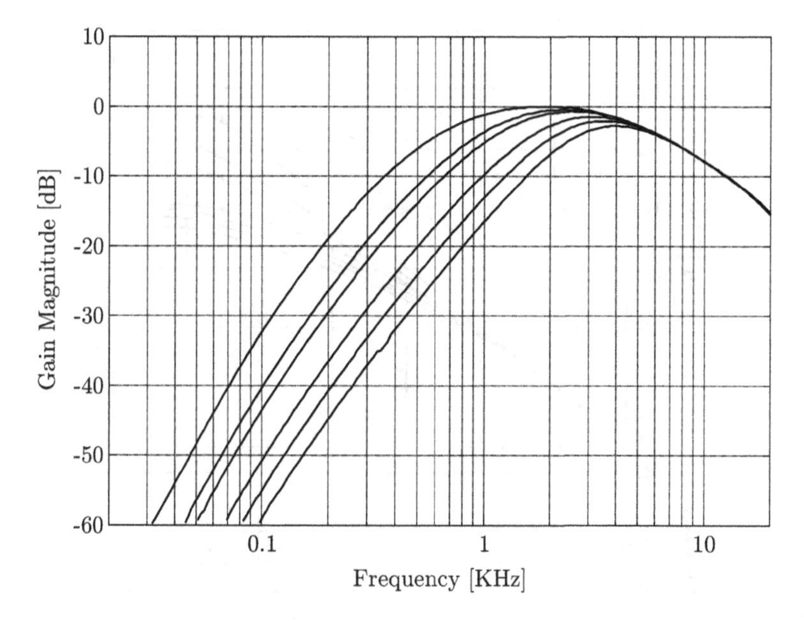

Figure 8.10. Experimental normalized transfer functions versus high-pass tuning.

any study about its yield. In the MOS context of this work, the main sources of sample-to-sample deviations are due to technology mismatching at the transistor level, as argued in Chapter 2. In this sense, it has already been demonstrated that such variations can be associated with threshold voltage deviations (ΔV_{TO}) between MOS transistors, even if they have been drawn following specific layout techniques.

At the system-on-chip level, ΔV_{TO} can be translated into three qualitatively different effects:

Gain errors. As a random variable in the V-domain, the constant value of ΔV_{TO} may be understood as a fixed gain programming (ΔV_{gain}) in the basic amplifying cell of Figure 3.3. Hence, yield can be limited for designs that include not only amplifiers or attenuators, but also unity-gain filters.

Frequency tuning deviations. The same threshold voltage random variations can cause deviations in the I-domain between drain currents. In case that such biasing levels play the role of tuning currents (I_{tun}) as studied in Chapter 4, their effect will be seen as a variability of the filter specifications (e.g. corner frequencies, quality factors, overshoots).

Output offsets. This last category belongs to a mixing between gain errors in the V-domain and full-scale bias deviations in the I-domain.

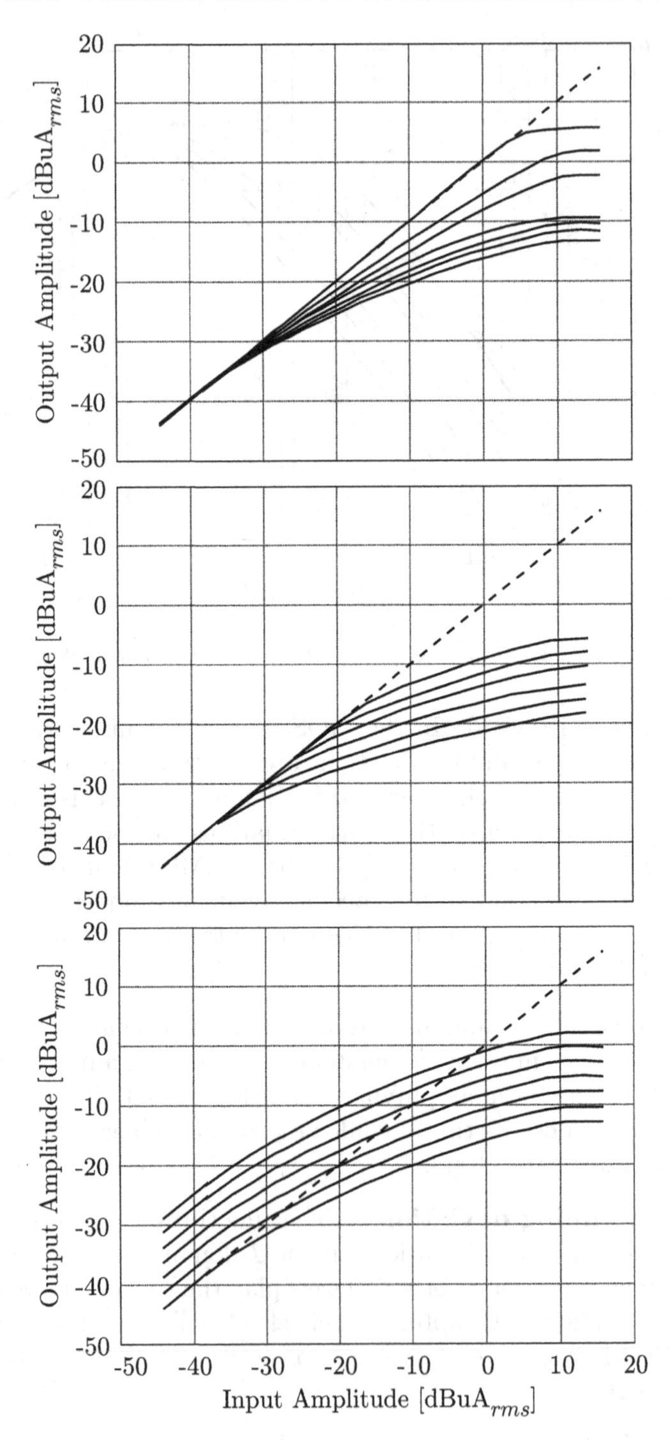

Figure 8.11. Experimental AGC steady-state input-output normalized response versus CR (upper), TK (middle) and G_{AGC} (lower) tuning.

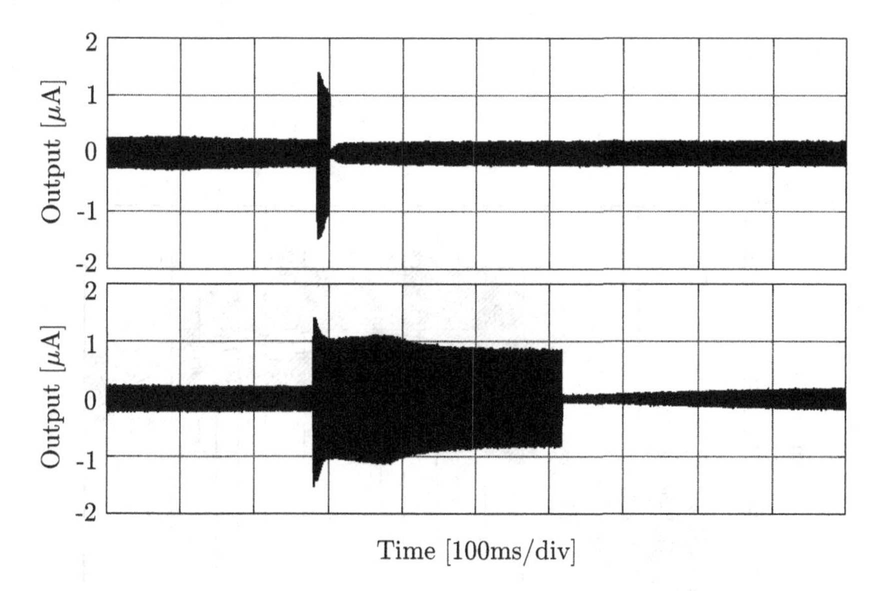

Figure 8.12. Experimental AGC output for a ±25dB input burst versus burst duration.

Typically, the combined effects of both errors cause non-full DC cancellation at the end of each stage, resulting in a non-null output offset.

In order to quantify all these non-idealities, some yield real data is presented in Figures 8.13 to 8.15 corresponding to the statistical analysis of 50 samples from the NEXO© pre-production. In the first graph, a frequency response window is shown for a given gain programming. Since no frequency-selective stages are enabled here and static output offsets are compensated during test, deviations caused by ΔV_{TO} are only due to gain errors. Secondly, Figure 8.14 presents a statistical window for a particular frequency-corner programming once compensated the gain-errors at low-frequency. Hence, the resulting deviations are also caused by ΔV_{TO} but through the tuning currents. Finally, last figure shows an input-output steady-state response window for a given compression ratio and threshold knee point code of the AGC stage. In this case, deviations due to ΔV_{TO} are a combined effect of both gain errors and output offsets at rectification. The overall yield information extracted from all these data is summarized in Table 8.2, where the reported values demonstrate the feasibility of this SoC for production, as well as the validity of the specific design methodology against technology mismatching proposed in Appendix A.

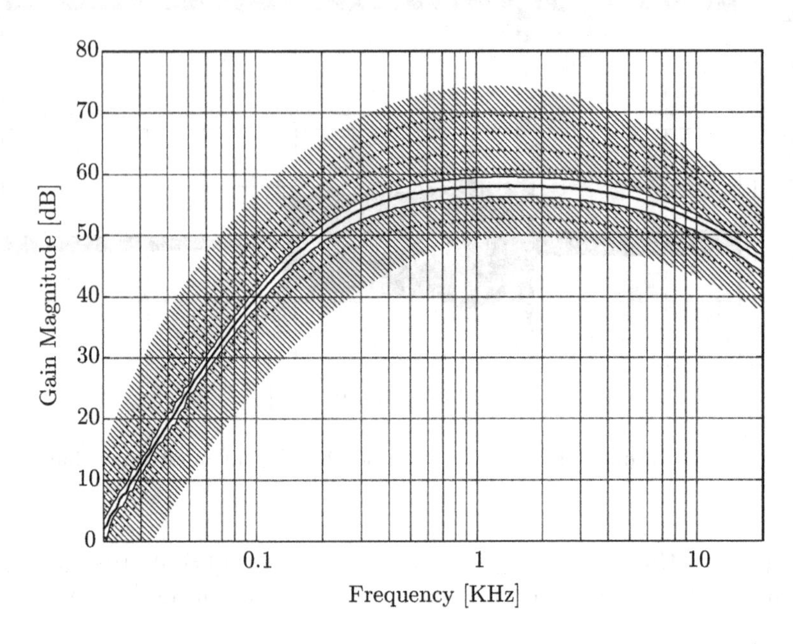

Figure 8.13. Example of experimental deviations ($\pm 2\sigma$ within striped areas) around an overall gain programming (solid) and other digitally programmable responses (dashed) for the AGC stage of the NEXO© system-on-chip.

Table 8.2. Summary of yield results at $\pm 2\sigma$ (i.e. 96% of samples) for the NEXO© system-on-chip.

Parameter	Value	Units
Gain Errors	1.7	dB
Frequency Tuning	3.5	%
Output Offset / Full Scale	8	%

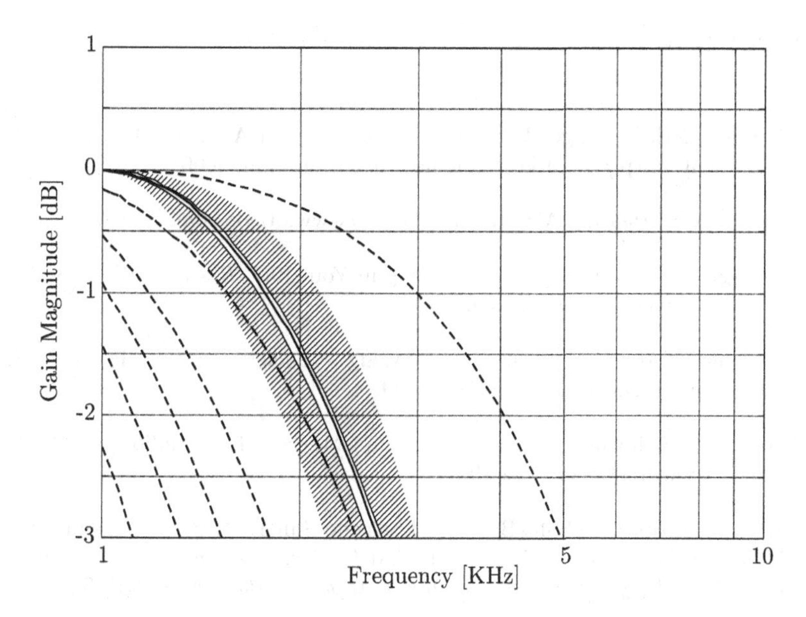

Figure 8.14. Example of experimental deviations ($\pm 2\sigma$ within striped areas) around the corner-frequency tuning programming (solid) and other digitally programmable responses (dashed) for the high-cut stage of the NEXO© system-on-chip.

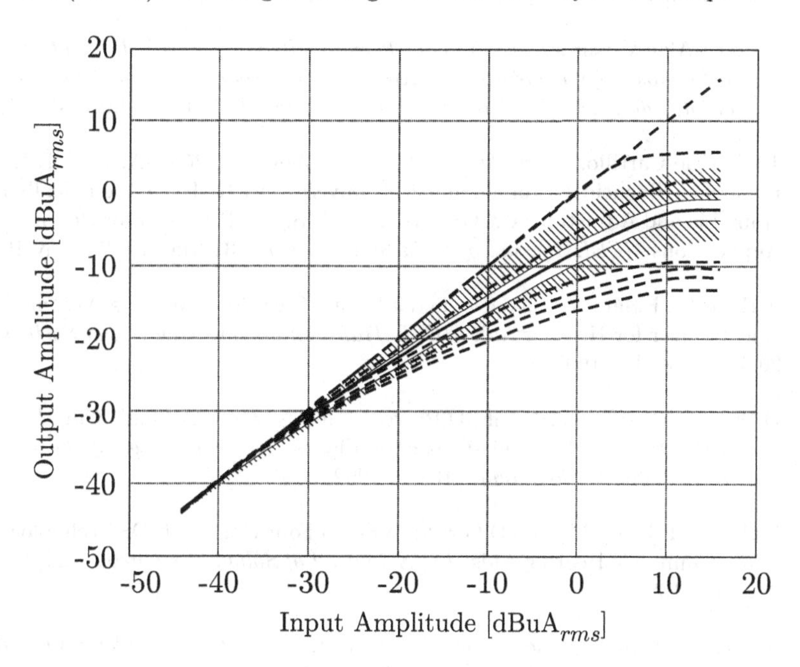

Figure 8.15. Example of experimental deviations ($\pm 2\sigma$ within striped areas) around a compression ratio programming (solid) and other digitally programmable responses (dashed) for the AGC stage of the NEXO© system-on-chip.

References

[1] The Kenneth W.Berger Hearing Aid Museum and Archives. Hearing Aid Development. http://dept.kent.edu/hearingaidmuseum, 2000.

[2] Oticon A/S. Hearing Aid History. http://www.oticon.com, Apr 1998.

[3] S.Baken and A.T.Palmer. A Microchip in Your Ear. *Business Week*, Aug 1999. http://www.businessweek.com.

[4] National Academy on an Aging Society. Hearing Loss. http://www.agingsociety.org, Dec 1999.

[5] Frost & Sullivan. The European Market for Audiology Products. http://www.frost.com, Aug 1997.

[6] J.Silva-Martínez, S.Solís-Bustos, J.Salcedo-Suñer, R.Rojas-Hernández, and M.Schellenberg. A CMOS Hearing Aid Device. *Journal of Analog Integrated Circuits and Signal Processing, Kluwer Academic Publishers*, 21:163–172, 1999.

[7] J.Silva-Martínez and J.Salcedo-Suñer. A CMOS Automatic Gain Control for Hearing Aid Devices. In *Proceedings of the International Symposium on Circuits and Systems*, volume I, pages 297–300. IEEE, 1998.

[8] R.Pérez-Aloe Valverde. *Técnicas de Diseño CMOS de Bajos Voltaje y Consumo con Reducidos Requerimientos de Area en Silicio: Aplicación a la Compensación de Pérdidas de Audición.* PhD thesis, Universidad de Extremadura, May 1996.

[9] J.F.Duque-Carrillo, Piero Malcovati, F.Maloberti, R.Pérez-Aloe, A.H.Reyes, E.Sánchez-Sinencio, G.Torelli, and J.M.Valverde. VERDI: An Acoustically Programable and Adjustable CMOS Mixed-Mode Signal Processor for Hearing Aids Applications. *IEEE Journal of Solid State Circuits*, 31(5):634–645, May 1996.

[10] A.H.Reyes, E.Sánchez-Sinencio, and J.Duque-Carrillo. A Wireless Volume Control Receiver for Hearing Aids. *IEEE Transactions on Circuits and Systems-II*, 20(1):16–23, Jan 1995.

[11] D.Wayne, M.Rives, T.Huynh, D.Preves, and J.Newton. A Single-Chip Hearing Aid with One Volt Switched-Capacitor Filters. In *Proceedings of the Custom Integrated Circuits Conference*. IEEE, 1992.

[12] F.Callias, F.H.Salchli, and D.Girard. A Set of Four IC's in CMOS Technology for a Programmable Hearing Aids. *IEEE Journal of Solid State Circuits*, 24(2):301–312, Apr 1989.

[13] Siemens Audiologische Technik GmbH. Private Communication. http://www.hoergeraete-siemens.de, Aug 1997.

[14] The MathWorks, Inc., 24 Prime Park Way, Natick, MA 01760-1500, USA. *Using MATLAB Version 5.2*, Jan 1998. http://www.mathworks.com.

[15] P.K.Ko and C.Hu. *BSIM3v3 Manual.* Department of Electrical Engineering and Computer Science, University of California, Berkeley, CA 94720, 1995. http://www-device.EECS.Berkeley.EDU/~bsim3.

[16] Y.Cheng, M.Jeng, Z.Liu, J.Huang, M.Chan, K.Chen, P.K.Ko, and C.Hu. A Physical and Scalable I-V Model in BSIM3v3 for Analog/Digital Circuit Simulation. *IEEE Transactions on Electron Devices*, 44(2):277–287, Feb 1997.

[17] Avant! Corporation, 46871 Bayside Parkway, Fremont, CA 94538, USA. *Star-Hspice Manual. Release 2000.2*, Jul 2000. http://www.avanticorp.com.

[18] Cadence Design Systems, Inc., 555 River Oaks Parkway, San Jose, CA 95134, USA. *Design Framework II Help 4.4.3*, Dec 1998. http://www.cadence.com.

[19] Microson-Gaes, 160, Pere IV, E-08005 Barcelona, Spain. *NEXO Products*, 2001. http://www.microson.es.

[20] Gennum Corp., 970 Fraser Drive, Burlington, Ontario L7L 5P5. *GA3201 Programmable DynamEQ II: Preliminary Datasheet*, Sep 1999. http://www.gennum.com.

[21] Oticon A/S, 58, Strandvejen, DK-2900 Hellerup, Denmark. *Swift 100: Technical Information*, Oct 2000. http://www.oticon.com.

[22] Siemens Audiologische Technik GmbH, D-91058 Erlangen, Germany. *Music: Technical Information*, 2001. http://www.hoergeraete-siemens.de.

Chapter 9

CONCLUSIONS

Abstract This last chapter can be understood as a summary of all the new knowledge generated by this work. The resulting conclusions range from the basic novel circuit techniques at the device level, to the proposed design methodologies for improving system design. Furthermore, future work tasks are also proposed in order to expand this research area in terms of signal processing, technology portability and dynamic range enhancement.

1. Results

Different conclusions can be summarized at the end of this research according to the initial motivation declared in Chapter 1: *the aim of this work is the research on novel analog circuit techniques based on the MOSFET operating in subthreshold to exploit the low-voltage capabilities of Log companding signal processing.* The main results are obtained from Chapters 3 to 7, concerning the basic research effort on new CMOS circuit techniques, and also from Chapter 8, for considerations at the system level in real applications. The collected conclusions are listed below following a bottom-top hierarchical scheme:

CMOS generalization of Log companding. This work demonstrates that the MOSFET is suitable to implement low-frequency Log processing. However, only some structures are allowed for exact synthesis at the transistor level. In this sense, the main contributions derived from the results are listed here:

- Gate- (GD) and source-driven (SD) Log companding functions based on the MOSFET operating in weak inversion.
- Basic topological restrictions and device matching requirements.

- Compatibility with non-separated wells (i.e. anti-latch-up rules in CMOS technologies).
- Generalization for amplification and AGC, arbitrary filtering, PTAT generation and PDM modulation.

Very low-voltage basic building blocks. Based on the above results, a wide collection of cells are proposed to implement the different types of signal processing studied. All the core devices of these basic building blocks operate in weak inversion. The novel features introduced at this level can be summarized as follows:

- Very low-voltage capabilities (down to 1V).
- Saturated and non-saturated topologies.
- Electronic and wide-range tuning (e.g. gain factors and corner frequencies).
- All auxiliary circuitry for frequency compensation, biasing, calibration and digital programming.

Design methodology for compact synthesis. Specific design procedures are presented for the above basic building blocks. Strategies range from general purpose stages to more complex systems, such as AGC loops and high-order filters. Particularly, the matrix procedure proposed for arbitrary filtering allows:

- Important Si area savings through circuit reductions.
- Optimized compressed operating points in terms of distortion.
- Very low-power consumption (few tens of μW).

Dynamic range versus power and area expressions. The common device-level equations concerning dynamic range issues are obtained. The role of the key design parameters power and Si area are identified:

- Total harmonic distortion due to moderate inversion degradation.
- Signal-to-noise ratio from thermal and flicker noise contributions.
- Procedure to design compressors and expanders.
- Comparison between different class operations.

All-MOS implementations. Taking advantage of the inner voltage compression in Log companding, the use of non-linear capacitors is proposed. The basic ideas presented in this field are listed as follows:

- General tuning-current compensation technique.
- The NMOS capacitor proposal.
- Integration through digital CMOS technologies.
- Si area savings or alternatively dynamic range improvements.

Design Techniques for very low-power audio systems-on-chip. All the proposed Log companding CMOS circuit techniques constitute an important set of design tools to be applied for the design of ASICs for low-frequency (up to 100KHz) applications. In particular:

- Experimental circuits operating at 1V for the different types of signal processing functions.
- First true 1V CMOS hearing-aid-on-a-chip without any charge-pump, and with one of the lowest current consumption levels of the market.

2. Future Work

The continuity of the work presented is illustrated by the following short- and mid-term activities:

Extension to very low-voltage A/D Conversion. Due to the market evolution described in Chapter 8, special efforts will be devoted to apply all the proposed CMOS techniques for the synthesis of very low-voltage audio A/D converters. In particular, oversampling $\Sigma\Delta$ topologies seem the best choice to take advantage of the new Filtering basic building blocks presented in Chapter 4. In this sense, some successful work has already been done [4].

All-MOS exact implementations. Further improvements of the all-MOS technique presented in this work may help to achieve the same distortion performances without the need of poly-Si capacitors. In that case, all the Log companding analog processing could be integrated in digital CMOS technologies, with the consequent compatibility with low-voltage DSP-based systems-on-chip and reduction in costs. What is more, these sub-micron technologies usually exhibit higher thickness reduction in gate oxide than in the equivalent poly-Si-poly-Si capacitance structure, as can be clearly seen in the capacitance densities C_{gate} and $C_{polypoly}$ of Table 9.1, respectively. Hence, the resulting increase in C_{gate} can be exploited to scale down the area of capacitors, with respect to the almost constant value of $C_{polypoly}$, or alternatively increase the power to achieve better dynamic range.

Table 9.1. Example of typical capacitance densities in CMOS technologies of a same foundry.

Parameter	Process A	Process B	Units
λ	1.2	0.35	μm
t_{oxgate}	18.5	7.5	nm
C_{gate}	1.87	4.6	fF/μm^2
$t_{oxpolypoly}$	50	40	nm
$C_{polypoly}$	0.69	0.86	fF/μm^2
$C_{gate}/C_{polypoly}$	2.7	5.4	-

Integration with NEMS. One of the future big challenges of any analog circuit technique is its compatibility with integrated nano-mechanical systems, either for signal transduction or even mixed (i.e. mechanical and electronic) signal processing. In this context, the technological compatibility of the new CMOS companding circuit techniques with low-frequency nano-mechanical sensors and actuators (e.g. Silicon microphone, planar receiver) should be investigated.

Dynamic biasing strategies. Apart from low-voltage capabilities, overall low-current consumption is also desired in system-on-chip applications. Hence, circuit techniques to implement the Class-H concept introduced in Chapter 7 may help to achieve larger dynamic range-to-power ratios at the system level.

References

[1] F.Serra-Graells. VLSI CMOS Low-Voltage Log Companding Filters. In *Proceedings of the International Symposium on Circuits and Systems*, volume I, pages 172–175. IEEE, May 2000.

[2] F.Serra-Graells. All-MOS Subthreshold Log Filters. In *Proceedings of the International Symposium on Circuits and Systems*, volume I, pages 137–140. IEEE, May 2001.

[3] F.Serra-Graells, L.Gómez, and O.Farrés. A True 1V CMOS Log-Domain Analog Hearing-Aid-on-a-Chip. In *Proceedings of the European Solid-State Circuits Conference*, pages 420–423. IEEE, 2001.

[4] F.Serra-Graells. 1V All-MOS $\Sigma\Delta$ A/D Converters in the Log-Domain. In *Proceedings of the International Symposium on Circuits and Systems*, volume II, pages 213–216. IEEE, May 2002.

References

1. [illegible]

2. [illegible]

3. [illegible]

4. [illegible]

Appendix A
Simulation and Test

Abstract This appendix should be considered as a set of tips and tricks for the practical verification and measurement of all the CMOS analog circuit techniques proposed in this work. The recipes given in the following sections cover from numerical simulation issues to the final experimental test. Practical examples are supplied for each item.

1. Numerical Simulation

When using CAD tools in the design of CMOS circuits and particularly for analog synthesis as in the context of this work, consistency of predictions is of particular importance in order to obtain satisfactory results. In this sense, the following issues should be checked:

1.1 SPICE Models

The need for accurate MOS transistor modeling in our Log companding environment has been already argued and introduced in Chapter 2. Nowadays, the de facto standard SPICE model supplied by the semiconductor industry for full-custom and analog design is BSIM3 [1, 2]. When properly extracted and fitted, this device model returns good enough accuracy for the circuit techniques proposed in this work.

However, special attention must be also paid to the particular SPICE-like simulator used. In some cases, the public BSIM3 code model has been only partially ported to the engine. Although the I/V large signal equations and the small signal parameters are always included, some other important parts like the intrinsic MOS capacitive expressions or the power spectral density (PSD) equations for noise may be taken from simpler models (e.g. BSIM1). This fact can be clearly seen in Figure 4.38, as part of the all-MOS filtering study presented in Chapter 4.

Also, technology mismatching is not usually included in the above SPICE models. Furthermore, the characteristic mismatching parameters A_{VTO} and A_β may not even be available at the foundry. In that case, the data can be extrapolated from Figures 2.8 and 2.9. Also, the simulation method proposed in Subsection 1.4 returns precise predictions for Gaussian distributions of any analysis.

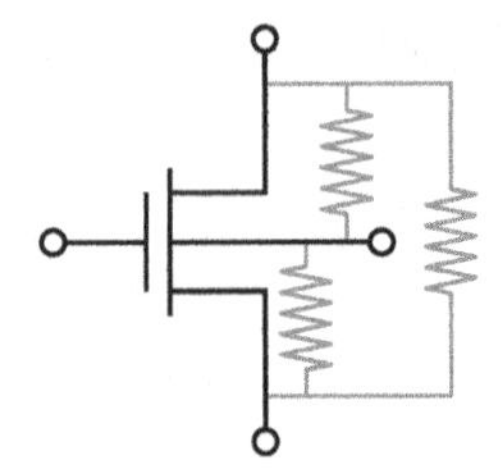

Figure A.1. Hspice convergence elements for the NMOSFET case.

1.2 Numerical Convergence

Numerical methods for solving non-linear circuit equations are usually optimized in many SPICE-like simulators for CPU-time consumption. Thus, it is the designer's responsibility to properly set the control parameters concerning accuracy and tolerance, which will determinate the final validity of any computed value.

For the companding frame of this work, special attention must be paid to resolution in the I-domain, requiring in some cases a change in the default control variables of the simulator to achieve the suitable current accuracy. The following items are illustrated for Hspice© [3], but can be easily translated to other SPICE-like simulators. All parameters are summarized in Table A.1.

Resolution. Typical default values for both absolute (e.g. `ABSVDC`) and relative (e.g. `RELVDC`) voltage accuracy are usually good enough to quantify internal compressed potentials in the V-domain. However, the equivalent parameters related to device currents do not usually cover the lower boundary of the I-domain dynamic range (DR). As a result, absolute (e.g. `ABSI` and `ABSMOS`) and relative (e.g. `RELI` and `RELMOS`) current accuracy must be set to suitable values according to I_{bias} levels.

Accuracy. Numerical methods for non-linear analysis (i.e. `DC` and `TRAN`) often introduce some linear elements to enhance circuit convergence. In the case of the MOS transistor, resistors are distributed as shown in Figure A.1 to improve continuity in moderate inversion and between conduction and saturation regions.

Although the internal compression in the V-domain tends to minimize the effect on branch currents, their conductance (e.g. `GMINDC` and `GMIN`) must be controlled according to the desired current resolution, and the maximum differential voltage drop possibly present in the circuit (i.e. V_{DD}). In any case, if the operating point convergence is difficult, the start-up approach should be followed: begin with a trivial zero-energy solution (i.e. all sources and charge in capacitors to null values), and perform a transient power-up to finally reach the desired operating point after relaxation. Such a final solution can then be used as initial conditions for any other non-linear analysis, either transient or static.

1.3 Large Signal Frequency Analysis

The spectral behaviour of circuits is usually simulated through small signal linear models (i.e. `AC`). However, second order effects related to signal amplitude itself tend to limit in practice the validity of the above approach.

Table A.1. Hspice simulation control values used in this work.

Parameter	Default	Recommended
ABSVDC	50μV	50μV
RELVDC	0.1%	0.1%
ABSI	1nA	1pA
ABSMOS	1μA	1pA
RELI	1%	1%
RELMOS	5%	1%
GMIN, GMINDC	1pS	0.1pS
PIVTOL	10^{-15}	10^{-15}

In the Log companding frame of this work, such an issue is of particular importance since signal processing is mainly based on large signal device I/V curves. In other words, design equations are directly supported by the non-linearity of the MOS device. Hence, practical problems may occur when moving between saturation and conduction, entering moderate inversion or reaching velocity saturation. As a result, some tool is necessary to include such signal amplitude dependencies into numerical spectral analysis. The simulation method chosen in Figure A.2 takes advantage of the transient impulse response [4] (i.e. `TRAN`) in order to compute the frequency transfer function through a discrete Fourier transform (DFT), more precisely using a standard fast Fourier transform (FFT) algorithm [5]. The basic advantage of the above procedure is the amplitude parameter (A_{in}), not available in small signal analysis. Then, symmetry (i.e. $\pm A_{in}$) and large signal (i.e. $A_{in} \to \infty$) effects can be explored and quantified. A practical simulation example for the all-MOS third-order low-pass filter of Section 7 is given in Figure A.3. Main drawbacks arise in the above procedure when requiring multi-decade frequency graphs. In these cases, the required number of points for the FFT makes the computation cost much more expensive than the simpler linear small signal simulation. A possible bypass for this situation consists of a spectrum splitting by calculating more than one impulse response at different time scales, as depicted in Figure A.4. The simulation-time savings obtained from this approach are illustrated in Figure A.5 for the same filter example with an input decoupling and $V \to I$ series network of $C = 47$nF and $R = 30$KΩ, respectively. Let us suppose a 6 decade spectrum analysis from 0.1Hz to 100KHz. Instead of computing a single $\delta(t)$ transient response of $10\text{s}/5\mu\text{s} \simeq 2^{21}$samples (i.e. >2Mpoint FFT!), two impulse responses of $10\text{s}/5\text{ms} \simeq 2^{11}$samples and $10\text{ms}/5\mu\text{s} \simeq 2^{11}$samples are performed (i.e. $2 \times$ 2Kpoint FFT). Obviously, final quantization of large signal range effects is preferably performed through fine steady state harmonic analysis (i.e. THD).

1.4 Technology Mismatching Simulation

Robustness in integrated circuits is as important as accuracy, so it must be considered during the design stage. In the particular case of CMOS processes, technological mismatching plays an important role for defining the absolute device areas in analog designs, such as for the Log companding circuit techniques of this work, as reported in Section 4. Hence, apart from process spread at run levels usually covered by the

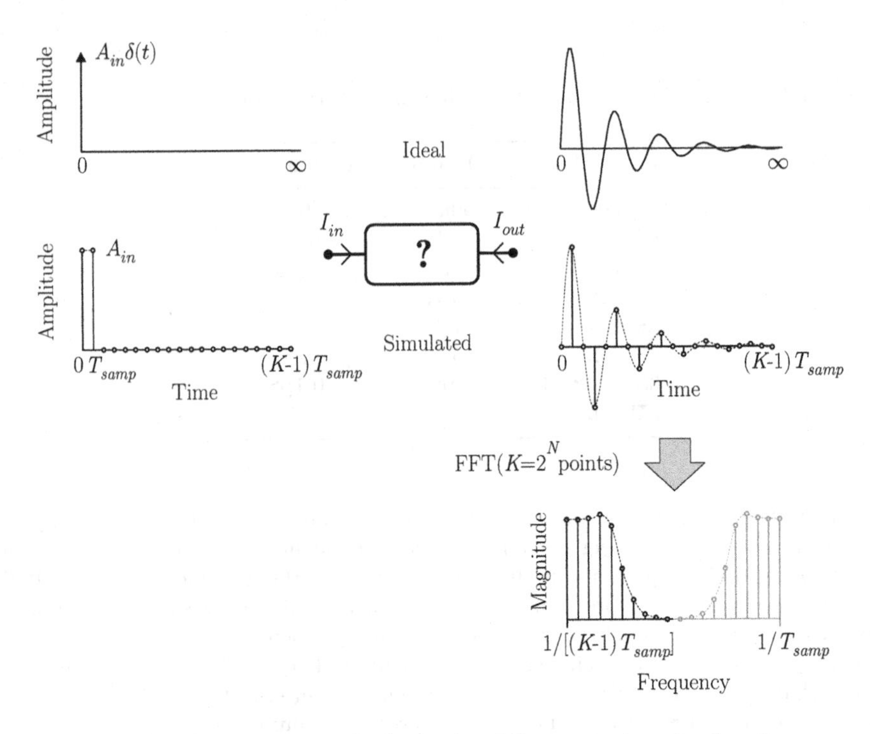

Figure A.2. Simulation strategy for large signal frequency transfer functions.

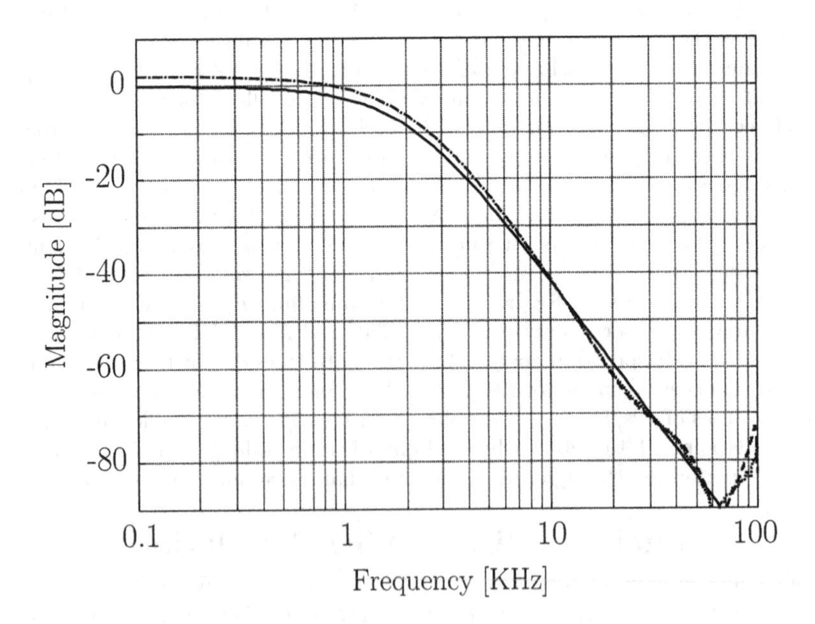

Figure A.3. Simulated small (solid) and large signal at $\pm 1\mu A_{peak}$ (dashed/dotted) sample frequency transfer functions.

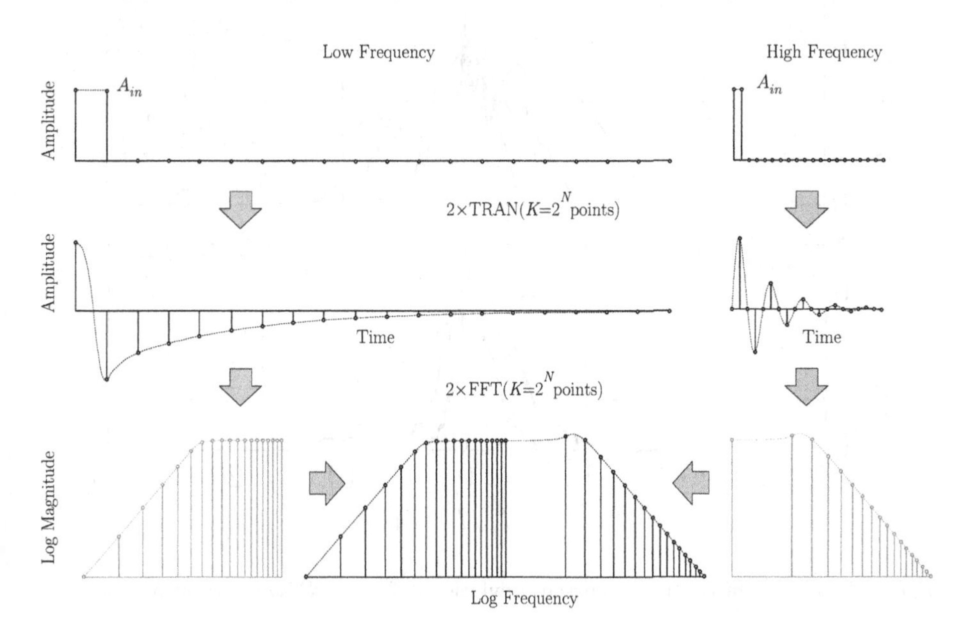

Figure A.4. Splitting strategy for multi-decade large signal frequency analysis.

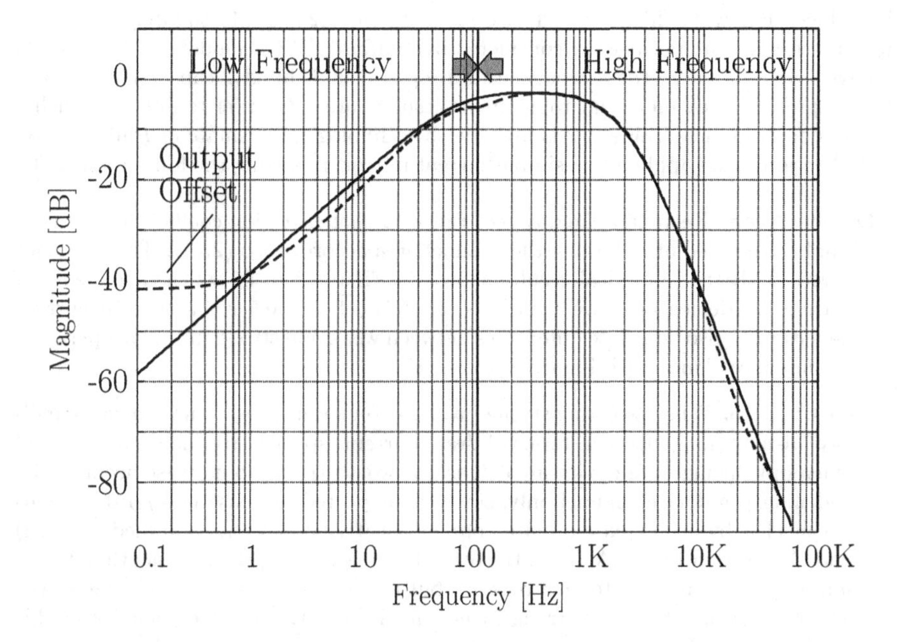

Figure A.5. Simulated small (solid) and large signal (dashed) 6 decade analysis.

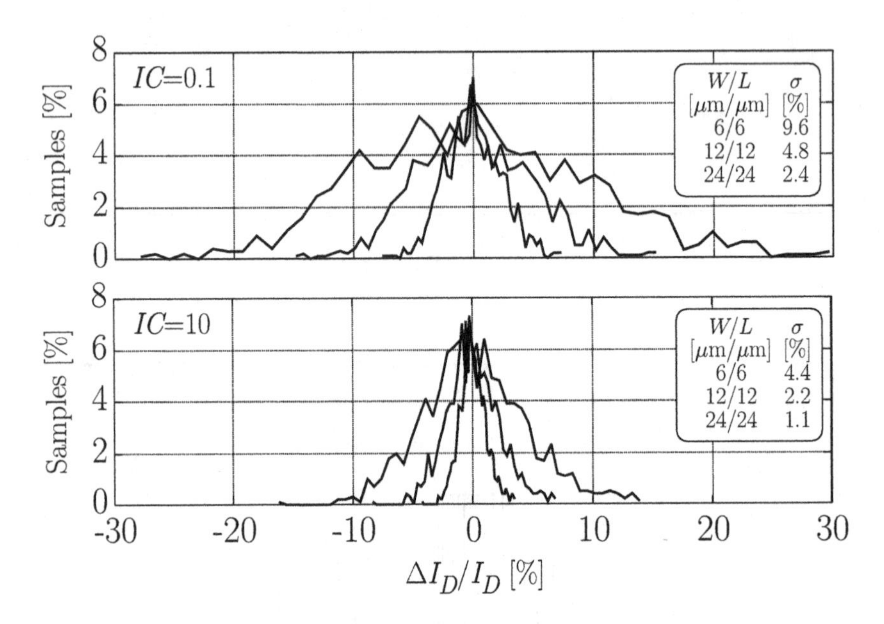

Figure A.6. Simulated output current deviations of a 1:1 NMOS current mirror.

corner models (e.g. slow versus fast or worse-speed versus worse-power cases), technological mismatching must also be taken into account during numerical simulation. Hence, local and Gaussian deviations for each MOSFET of the schematic should be introduced following the general device area rule in (2.24). Unfortunately, such geometrical models are not usually considered in standard SPICE models for Montecarlo analysis. In this sense, the new algorithm proposed in Table A.2 can be used to perform a precise study on circuit robustness versus Si area. In order to illustrate such a simulation approach, Table A.2 is fitted with technological mismatching information from Section 5 and applied to a simple current mirror, reporting the following results:

Device Area. Keeping a fixed aspect ratio, so that inversion coefficient, accuracy can be improved by enlarging the channel area according to (2.24). This behavior agrees with the numerical results shown in Figure A.6, in which increments of 4 times in device area are translated to $\times 1/\sqrt{4} = \times 0.5$ reductions in current deviations. Also, absolute differences between weak and strong inversion operation are in concordance with Figure 2.7.

Aspect ratio. When selecting strong inversion region in order to minimize technology mismatching effects in terms of drain current, special attention must be paid when optimizing final device area. It is not worth trying to increase accuracy by widening the device channel only, since the expected decrease in V_{TO} deviations is canceled by an increase of gate voltage sensitivity to V_{TO}, as argued in (2.28) and observed in Figure A.7. On the other hand, enlarging channel length not only improves accuracy due to area reasons, but also due to the fact of pushing up the gate bias point at deeper strong inversion. However, the direct penalty in this case is a degradation of low-voltage compatibility due to higher voltage drops.

Table A.2. Hspice Montecarlo simulation of V_{TO} local mismatching.

---

0 **Require** mismatching coefficient <AVTHO>

1 **Substitute** each MOSFET element:

```
M1 <D> <G> <S> <B> <modelname> L=<l> W=<w> M=<m> ...
```

by a subcircuit call:

```
XM1 <D> <G> <S> <B> <modelname> PL=<l> PW=<w> PM=<m> ...
```

2 **Replace** the common typical modelcard:

```
.MODEL <modelname> NMOS LEVEL=<level> VTHO=<typvto> ...
```

by a subcircuit definition with scalable Gaussian distributions:

```
.SUBCKT <modelname> <D> <G> <S> <B>
M1 <D> <G> <S> <B> MONTEMOD L=PL W=PW M=PM ...
.PARAM MCVTHO=AGAUSS(<typvto>,'<AVTHO>/sqrt(PL*PW*PM)',1)
.MODEL MONTEMOD NMOS LEVEL=<level> VTHO=MCVTHO ...
.ENDS <modelname>
```

3 **Perform** a standard Montecarlo analysis.

---

2. Experimental Test Setup

The Log companding nature of the new CMOS circuit techniques presented in this work also introduces some specific lab necessities for their experimental measurement. In particular, the following requirements should be satisfied:

Current mode for both the in-going and out-coming I-domain signals of the device under test (DUT). Since the standard electrical test is usually based on voltage measurements, an input $V \rightarrow I$ and output $I \rightarrow V$ conversion stages must be included in the test setup. Practical full-scale signal specifications for all the proposed circuit techniques are located below $< 5\mu A_{peak}$. Also, front-end output and back-end input impedances in current-mode should be $> 10K\Omega$ and $< 100K\Omega$, respectively.

Low-frequency spectrum analysis. As argued in Chapter 1, the new CMOS design strategies are suitable for voice and audio applications. Hence, front and back-end

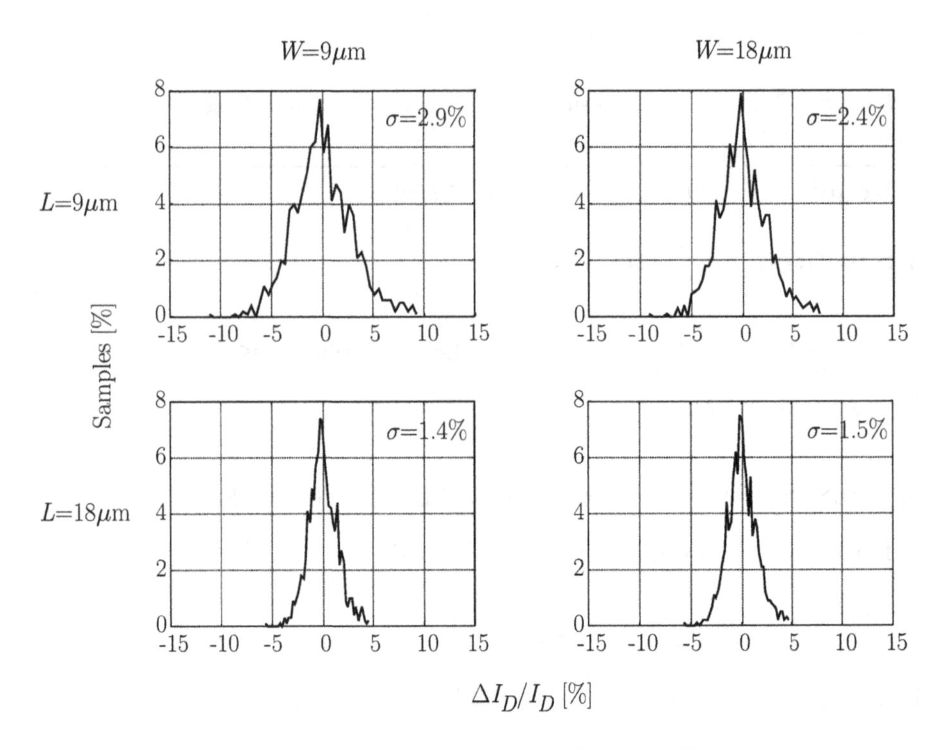

Figure A.7. Simulated output current deviations of a 1:1 NMOS current mirror operating in strong inversion (IC=1μA/100nA) versus channel aspect ratio.

instrumentation with a bandwidth of up to 100KHz is large enough for harmonic analysis. Furthermore, it is preferable to have a programmable low-frequency high-order low-pass filter available at the backend for noise measurement.

Large dynamic range capabilities which should be negligible compared to the DUT. In this sense, practical circuit performances in the frame of this work are around $0.1\% < THD < 1\%$ and $60\text{dB} < DR < 80\text{dB}$.

Taking into account all the above specifications, the general lab setup proposal is depicted in Figure A.8. The DUT operates by means of a local power supply in order to avoid noise and spurious signals from any external power line. Such a voltage supply can be selected from a single cell battery (i.e. $V_{DD,SS} \simeq \pm 0.6\text{V}$) or from the programmable dual regulator RC4194© [6] (i.e. $V_{DD,SS} > \pm 50\text{mV}$). Stimuli are synthesized by the low-distortion function generator DS360© [7] in conjunction with a series resistor $R_{conv} = 100\text{K}\Omega$ to perform the $V \rightarrow I$ current conversion. Also, a capacitor $C_{dec} = 22\text{nF}$ is inserted to decouple DC bias levels. The resulting sensitivity is 1μA/0.1V above 75Hz. Once processed by the DUT, the outcoming signal is converted back from $I \rightarrow V$ using the very low-noise current preamplifier LCA400K10M© [8] with an equivalent transresistance of 10V/1μA. Since the linear range of such an amplifier is limited to $\pm 1\mu$A, an alternative equipment for larger amplitudes is the SR570© box [7], which enables different scale conversion but also exhibits less dynamic range (other preamplifiers for photodetectors can also be used). The resulting waveform is captured by the dynamic signal analyzer SR785© [7] for

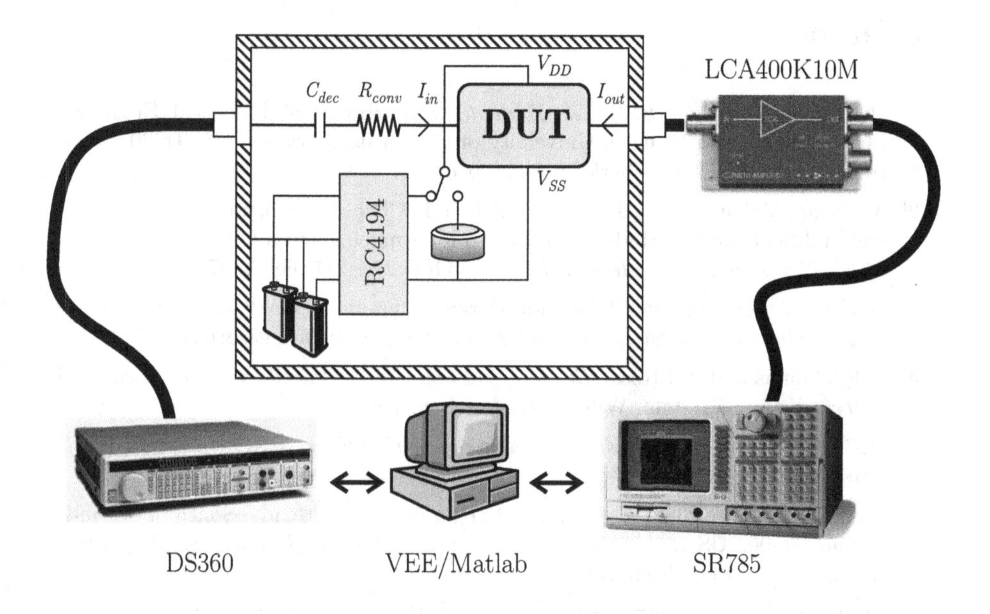

Figure A.8. General test setup proposal.

Table A.3. Equivalent noise (I_{noise}) and total harmonic distortion (THD) contributions in Figure A.8. noise is computed from 100Hz to 10KHz either at the DUT input or output (unity gain case). Supply voltage noise due to RC4194 is about 250μV$_{rms}$.

Parameter	DS360	R_{conv}	DUT	LCA400K10M	SR785	Units
I_{noise}	0.01	0.04	> 0.1	0.007	0.0001	nA$_{rms}$
THD	0.001	n.a.	> 0.1	n.a.	0.01	%

the transfer function, noise or distortion analysis. The overall guarding is guaranteed through BNC connectors and a shielded box for both the DUT and local supplies, as drawn in Figure A.8. Also, signal integrity along the path can be verified in Table A.3. Finally, automatic measurements may be programmed through PC-based control software such as VEE© [9] and Matlab© [10].

References

[1] P.K.Ko and C.Hu. *BSIM3v3 Manual.* Department of Electrical Engineering and Computer Science, University of California, Berkeley, CA 94720, 1995. http://www-device.eecs.berkeley.edu/~bsim3.

[2] Y.Cheng, M.Jeng, Z.Liu, J.Huang, M.Chan, K.Chen, P.K.Ko, and C.Hu. A Physical and Scalable I-V Model in BSIM3v3 for Analog/Digital Circuit Simulation. *IEEE Transactions on Electron Devices*, 44(2):277–287, Feb 1997.

[3] Avant! Corporation, 46871 Bayside Parkway, Fremont, CA 94538, USA. *Star-Hspice Manual. Release 2000.2*, Jul 2000. http://www.avanticorp.com.

[4] R.E.Thomas and A.J.Rosa. *Circuits and Signals: An Introduction to Linear and Interface Circuits.* John Wiley and Sons Inc., 1984.

[5] R.C.Dorf, editor. *The Electrical Engineering Handbook.* CRC Press and IEEE Press, 1997.

[6] Fairchild Semiconductor Corp., 82 Running Hill Road, South Portland, Maine 04106, USA. *RC4194 Dual Tracking Voltage Regulators*, May 1998. http://www.fairchildsemi.com.

[7] Standford Research Systems Corp., 1290-D Reamwood Avenue, Sunnyvale, CA 94089, USA. *SRS Catalog*, 1998. http://www.thinksrs.com.

[8] FEMTO Messtechnik GmbH., Stargarder Strasse 74, D-10437 Berlin, Germany. *LCA-400K-10M Datasheet*, Jul 1999. http://www.femto.de.

[9] Agilent Technologies Inc., 395 Page Mill Road, Palo Alto, CA 94306, USA. *VEE OneLab User's Guide*, Mar 2000. http://www.agilent.com.

[10] The MathWorks, Inc., 24 Prime Park Way, Natick, MA 01760-1500, USA. *Using MATLAB Version 5.2*, Jan 1998. http://www.mathworks.com.